PERIGLACIAL MASS-WASTING:
A REVIEW OF RESEARCH

by

Charles Harris B.Sc.,Ph.D.
Geography Section
Department of Geology
University College Cardiff

Geo Abstracts
(Geo Books)
NORWICH

Published by Geo Abstracts, Norwich

British Geomorphological Research Group, Research
Monograph Series, 4

ISBN 0 86094 078 0

Also in this series:
No. 1. K. Gilman & M.D. Newson, Soil pipes and pipeflow.
       A hydrological study in upland Wales. 1980.
No. 2. J.M. Verstraten, Water-rock interactions. A case
       study in a very low grade metamorphic shale catch-
       ment in the Ardennes, NW Luxembourg. 1980.

       (Paperback with Appendix on microfiche, or
        Hardback complete with Appendix, pp 177-244)
No. 3. J.A.A. Jones, The nature of soil piping: a review
       of research. 1981.

Other titles in preparation

Geo Abstracts, University of East Anglia, Norwich NR4 7TJ,
England

Printed in Great Britain by Headley Brothers Ltd The Invicta Press Ashford Kent and London from artwork supplied

AUTHORS' PREFACE

The importance of mass-wasting processes in the periglacial zone is widely acknowledged. Mass-wasting is of interest not only to the geomorphologist, but also to the engineer and those concerned with land-use and development in cold-climate regions. Research by physicists, engineers, geologists and geomorphologists into frozen ground phenomena has accelerated over the past ten or fifteen years and now provides a considerable fund of basic knowledge. This monograph attempts to bring together some of this research in a review of periglacial mass-wasting processes, their resulting sediments, and the landforms they produce.

Mass-wasting is taken to include those processes causing downslope mass displacement of sediment in the active layer, generally associated with thawing soil ice. Excluded therefore are: deep seated landslides, which penetrate below the level of seasonal freezing and thawing; movements of sediment due to slush and snow avalanches; rock glaciers, which may move as a result of deformation of interstitial ice rather than as a result of active-layer thawing.

The approach has been to emphasise recent work on the physics, and geotechnical properties of unconsolidated sediments in the active layer, in order to show the relevance of this kind of information to geomorphological research. This monograph is therefore complementary in many respects to the textbooks on periglacial geomorphology that have been published in the 1970's in that it examines a narrower range of topics in greater depth and with a particular emphasis.

Following the introductory first chapter, chapter 2 considers the ground thermal regime, particularly factors influencing soil freezing and thawing. In chapter 3 soil freezing mechanisms are discussed together with resulting frost heave and soil creep. The analysis of slope stability in periglacial areas is reviewed in chapter 4 and the theory of thaw consolidation is introduced. Chapter 5 deals with solifluction processes, while chapter 6 describes more rapid types of periglacial mass movements. Thus in the first six chapters emphasis is placed squarely on the need for detailed measurement of the environmental conditions and the geotechnical and geophysical properties of sediments in studies of present day periglacial mass-wasting processes.

Finally, in chapter 7, the interpretation of fossil periglacial slope deposits by reference to modern analogues is illustrated in a review of periglacial slope deposits in Britain. Once again detailed sedimentological and geotechnical analysis of these deposits are required to enable quantitative comparisons to be made between them and sediments in the modern periglacial zone.

My initial interest in periglacial geomorphology arose
through the University of Reading Okstindan Research
Project, and I wish to extend my thanks to Peter Worsley
and the other members of Okstindan field parties for
providing the opportunity to do research in a periglacial
environment. Thanks are due to Dr. John Matthews and
Dr. Edward Watson who kindly read and commented on parts
of the manuscript. Several individuals have provided
previously unpublished material and these are acknowledged
in the text. Miss Louise Whittle prepared most of the
diagrams for final drawing and typed the manuscript.

Charles Harris

August 1980

CONTENTS

1. PERIGLACIAL MASS-WASTING, AN INTRODUCTION

Mass-wasting processes: the problem of terminology

Washburn (1979), following Flint and Skinner (1977) defines mass-wasting as 'the movement of regolith down-slope by gravity without the aid of a stream, a glacier or wind'. Mass movements on slopes in periglacial regions are often referred to as 'solifluction', following Andersson's original definition 'the slow flowing from higher to lower ground of masses of waste saturated with water' (Andersson 1906, p.95). A considerable degree of confusion however has resulted from the application of this general term to a range of processes. Three distinct problems have arisen: firstly the climatic conditions under which solifluction occurs; secondly the range of movement processes which may be included; and thirdly the rates of movement to which the term may be applied.

Considering the first of these problems, Andersson in his definition did not specify any one type of climate under which solifluction might occur, but suggested that it was most active under periglacial conditions. Solifluction has come to be associated with cold climates since Andersson's original definition, and Brown and Kupsch (1974) in their review of permafrost terminology regard solifluction as 'applicable to cold climates only, although the process is not confined to permafrost regions' (Brown and Kupsch 1974, p.36). Others including Dylik (1951, 1967), Baulig (1956, 1957) and Washburn (1967), have maintained that since Andersson did not specify periglacial environments in his original definition, no such climatic restriction on the use of the term solifluction should be made. As Dylik (1951, 1967) points out, Andersson suggested that the water which saturates the waste mantle may be derived from snow melting or rain and no reference was made to ground ice or frozen ground in Andersson's original definition.

In order to overcome the problem of applying a general term such as solifluction to specifically periglacial mass movements Dylik (1951) suggested the word 'congelifluction' to refer to 'earth flow occurring under conditions of frozen ground' (p.6). Baulig (1956, 1957) however, coined the term 'gelifluction' to be applied to soil flow associated with frozen ground irrespective of whether permafrost is present, and Washburn (1967) adopted this term, suggesting that the distinction between the presence or absence of permafrost may be more theoretical than practical, especially where there is deep freezing but no permafrost. In recent geomorphological literature the term gelifluction is preferred to solifluction when referring to saturated soil flow associated with soil freezing (e.g. Washburn 1973, Embleton and King 1975, and French 1976), although the term solifluction is retained by some, including Williams (1966), Dutkiewicz (1967), Benedict (1970), Harris (1972, 1977), Price (1973), and

McRoberts and Morgenstern (1974).

Since virtually all the literature on this type of mass movement refers to the periglacial zone, objections to the use of the term solifluction on the grounds of its lack of a specific climatic connotation appear of little practical relevance. Certainly solifluction is a far more important geomorphological process in periglacial areas than it is elsewhere.

A second source of confusion over terminology is the range of processes to which the term solifluction may be applied. It is generally agreed that three major processes may be operative in causing downslope mass movements in periglacial areas, namely soil flow, frost creep, and the sliding of largely intact units of soil over discrete shear planes. These processes may operate individually or in combination at any particular site.

Soil creep was defined by Sharpe (1938, p.21) as the 'slow downslope movement of superficial soil or rock debris, usually imperceptible except to observations of long duration', and he included creep with solifluction in a class of movement called 'soil flowage'. However, Davidson (1889) described the process whereby soil may creep downslope under periglacial conditions as resulting from frost heave lifting soil particles in a direction perpendicular to the ground surface, followed by a more nearly vertical resettling during thaw, under the influence of gravity. This mechanism of frost creep does not involve saturated flow of the soil. The basic difference between creep and flow has been widely recognised (Taber 1943, Sigafoos and Hopkins 1952, Williams 1957, Jahn 1961, Dylik 1967, Washburn, 1967, Benedict 1970, Harris 1972a, French 1976) although it is difficult if not impossible to distinguish between the two processes in the field where they operate together. McRoberts and Morgenstern (1974) stress this difficulty, and suggest that attempts to separate flow and creep are consequently of little practical relevance. French (1976) suggests that the term gelifluction should describe the slow flow of saturated soil under periglacial conditions while solifluction be used to indicate downslope soil movement resulting from the combined operation of soil flow and frost creep.

A potentially more serious source of confusion is the application of the term solifluction to translational slides developed in permafrost environments. Shallow slab slides in stiff clays have been observed in gentle slopes in south east England (Skempton and Petley 1967, Weeks 1969, Chandler 1970) and interpreted as periglacial mass movements initiated under permafrost conditions during the Plestocene. The surface morphology of some of these slides resembles that produced by gelifluction (lobes and terraces), leading Weeks (1969) to refer to them as 'solifluction lobes' and 'solifluction sheets'.

The slides consist generally of a mantle of stiff clayey mudslide debris above slip surfaces formed by continuous shears running sub-parallel to the ground surface, plus discontinuous shears which tend gradually to approach the ground surface when traced down slope, and referred to as emergent shears (Hutchinson 1974). Hutchinson suggests a process of sliding during periods of high pore pressure in the active layer as it thaws. Most of the shearing is along discrete shear planes, with relatively little disturbance of the moving mass itself, and Hutchinson refers to this process as 'periglacial solifluxion'. However, this sliding of shallow slabs of clay does not resemble the process of saturated soil flow described by Andersson in his original definition of solifluction.

The use of a term with genetic connotations to describe features with differing genesis, albeit developed under the same environmental conditions, may clearly lead to uncertainties and confusion. It would appear preferable to restrict the term solifluction to soil flow processes and to adopt different terminology for slides (Dylik 1967, Carson 1978). Periglacial mudslides (Chandler 1972) or periglacial slabslides (Chandler 1970) might adequately describe this type of mass movement and its resulting landforms.

The third source of disagreement over the use of the term solifluction lies in its application to soil movements of different rates, from slow imperceptible flow to rapid relatively fluid displacements. Högbom (1914) and Cailleux and Tricart (1950), use the term solifluction to include both slow creep and rapid mud flows although, Andersson's definition specifies slow flow.

Sigafoos and Hopkins (1952) differentiate between slow and rapid viscous flow, and Siple (1952) refers to slow flow as solifluction and rapid flow as downhill slump. Similarly Lamothe and St. Onge (1961) and Lundqvist (1962) identify slow flow as solifluction and more rapid flows with higher moisture contents as mud flows. Rapp (1962) refers to short duration rapid slope failures as 'momentary mass movements', as distinct from slow displacements which he calls 'continuous mass movements', including solifluction under the latter heading. Dylik (1967) points out that a continuum exists from relatively dry, very slow creep displacements, to water-dominated very rapid sheet wash, and intermediate stages include solifluction which is wetter and faster than creep, and mud flow, which is wetter and faster than solifluction.

McRoberts and Morgenstern (1974), following the usage established in geotechnical practice by Varnes (1958) divide landslides into general categories of flow, slide and fall dominated mass movements. Their field studies in Northern Canada indicate that landslide forms associated with the thawing of permafrost slopes could be described as being flow dominated, and could be subdivided into solifluction, skin flows, and bimodal flows. The main criteria for this subdivision are speed of movement and surface form. Solifluction is a relatively slow flow affecting the whole or part of the active layer, skin flows are shallow more rapid flows, often ribbon-like in form, involving the detachment of a thin veneer of vegetation and mineral soil. McRoberts (1978) suggests that solifluction might be considered to be the characteristic deforming mode of many naturally occurring active layers, but skin flows usually occur as a result of catastrophic events such as forest fires, heavy rains or high temperatures which increase the rate and depth of thaw. Bimodal flows develop where rapid thawing of permafrost occurs on steep scarp-like slopes, and consist of a steep upper section where permafrost thawing releases sediment which falls or flows to the foot of the scarp where it spreads out as a mudflow apron or lobe.

There is therefore widespread agreement that solifluction refers to slow saturated soil flow, and the term should not be applied to rapid flows of relatively short duration.

The discussion above illustrates the wide and often conflicting usage of the term solifluction. It has become necessary to re-define the term or adopt alternatives where specific processes are to be inferred. Periglacial mass movements are classified in table 1 below.

Table 1 Classification of periglacial mass movements.

Rate of Movement	Nature of Movement		
	Flow-type	Intermediate	Slide
Very slow (measured in mm/yr or less)	Frost Creep		
Slow (measured in cm/yr)	Gelifluction		
Fast (measured in m/yr - m/day)	Skinflow Bimodal flow		Slab slide Active layer glide

The terminology shown in table 1 will be used in this account, with the addition of 'solifluction', which will be used to describe the slow downslope flow of water saturated sediments due to the combined effect of gelifluction and frost creep. The term solifluction is particularly useful in describing topographic features, such as solifluction lobes, terraces and sheets, where the relative importance of gelifluction and frost creep is unknown.

The general term 'periglacial mass-wasting' will be taken to refer to downslope mass movements which are directly related to the thawing of frozen ground.

The periglacial environment

In 1906 J.G. Andersson reported on his studies in Bear Island and described a cold non-glacial environment where slow saturated mass-wasting was the dominant slope process. This characteristic form of mass movement he termed solifluction and the cold climatic environment in which it occurred he called 'subglacial'. Subsequently a Polish geologist, W. Lozinski introduced an alternative to the obviously ambiguous subglacial when he described the cold climatic zone immediately adjacent to the Pleistocene ice sheets as 'periglacial' (literally 'around the glacier'). Since its introduction 'periglacial' has been widely used to describe cold climatic areas irrespective of their proximity to glaciers, and has replaced Andersson's earlier term.

Lozinski's original work was concerned with frost shattering in the Carpathian Mountains (Lozinski 1912) and this process he considered to be most important in the periglacial zone. As more studies have been made in modern periglacial areas many other geomorphological processes have been recognised as either operating only under periglacial conditions or being particularly active under such conditions. In the latter category may be included the various forms of mass wasting (Dylik 1964).

Definition of the periglacial zone in climatic terms is difficult because of the many criteria which may be taken as characterizing periglacial areas. For instance, Péwé (1969) considers that the modern periglacial zone may be defined by the presence of permafrost, yet much of the Alpine tundra of mid-latitude mountains lacks permafrost, periglacial mass movements here being associated with the thawing of seasonally frozen ground. French (1976) suggests that periglacial environments may be defined as those in which frost action processes dominate, and he emphasises that there is no perfect spatial correlation between areas of intense frost action and areas underlain by permafrost.

Periglacial climates

As Price (1972) points out, there is as much difference between the contrasting climates of the periglacial zone as between those of the middle latitudes, and some subdivision is therefore necessary.

Tricart (1970) has suggested a classification of

climates based on the work of Troll (1944, 1958). Three climatic types are identified, on the basis of temperature and precipitation. Dry climates with severe winters occur in Arctic continental areas such as Siberia and northern Canada and have extremely low winter temperatures, short summers and low precipitation. Severe cooling of the ground results in permafrost being ubiquious.

The second climatic type, humid climates with severe winters is subdivided into Arctic and mountain subtypes. The former is typical of the coastal fringes of Alaska and Siberia, and Spitsbergen. It has higher precipitation and a smaller annual range of temperatures than the dry climates of the continental interiors. Permafrost is discontinuous. The mountain subtype occupies the mid-latitude uplands where winter temperatures are less severe than in the Arctic subtype, precipitation is much greater and permafrost generally absent.

The third type, climates with small annual temperature range, is also subdivided, into island climates and mountain climates of low latitudes. The island climates show small annual range (around $10^{0}C$), but relatively large short term temperature fluctuations associated with the passage of different weather types. Precipitation is usually moderate, permafrost is absent, and freezing of the ground occurs over periods of several days rather than seasonally. The mountain climates suffer low temperatures due to altitude, and seasonal variation, as in most tropical areas, is small. However, the high angle of the sun produces large diurnal temperature ranges and diurnal freezing and thawing of the ground, rather than permafrost. Precipitation may be high, so that running water is often the dominant denudational agent.

French (1976) modifies Tricart's classification, basing his classification on temperatures alone, and his summary data are reproduced in table 2. Vincocaya, in the Andes, at latitude $15^{0}S$ is not included in French's original table, but illustrates the climate of low latitude high altitude regions.

The High Arctic climates have a strong seasonal pattern of temperature fluctuation, but a weak diurnal pattern. Despite a fairly wide range in precipitation totals all areas are characterised by wet ground conditions during summer, due to the low rates of evaporation, and continuous permafrost.

Continental climates show a greater temperature range than in the High Arctic, with severe winter temperatures, but remarkably high temperatures in summer. Precipitation amounts are also higher than in the High Arctic zone, but with high summer evaporation rates the ground is usually drier. Permafrost is generally discontinuous. The boundary between these two climatic regions may for practical purposes be taken as the tree line, so that the

Table 2 - Periglacial climatic data

		Mean annual temperature	Annual range	Total precipitation
		°C	°C	mm
High Arctic	Spitsbergen	-8	25	298
	Sachs Harbour	-14	36	93
Continental	Yakutsk	-10	62	247
	Dawson City	-5	45	343
Alpine	Sonnblick, Alps	-7	15	1638
	Niwot Ridge, Rockies	-3	22	1021
Low temperature	Jan Mayen	0	8	365
range	South Georgia	+2	7	1309
	Vincocaya (4380m) Andes	+1.9	6	263

High Arctic climate prevails in the Arctic tundra vegetational zone, and the continental climate in the forested zone. In North America the boundary between continuous and discontinuous permafrost is also close to the tree line (Mackay 1972) and in the U.S.S.R., east of the Urals it coincides with the northern limit of the coniferous forests (Baranov 1959).

The Alpine climate corresponds to the mountain subtype of Tricart's humid climatic zone. It is characterised by less severe winter temperatures than the two previous areas, but well developed seasonal and diurnal temperature fluctuations. Precipitation is often high, much of it occurring as snow, and permafrost is usually lacking or discontinuous. Botanically the Alpine zone lies above the tree line (Dahl 1956, Rune 1965) and this provides a convenient boundary for the Alpine climatic zone. Price (1973) points out that both the arctic and the alpine tree lines generally parallel the 10°C isotherm for the warmest month.

The fourth climatic type in French's classification, climates of low annual temperature range, corresponds to Tricart's third climate type, and is similarly subdivided into island climates and mountain climates in low latitudes.

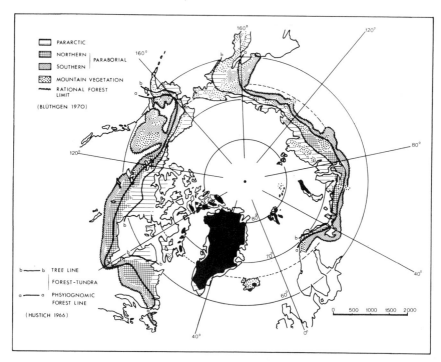

Figure 1 The subarctic, according to Blüthgen (1970) and Hustich (1966).

Of French's four periglacial climatic zones, only the continental climate lies below the tree line. The other three climates may be grouped together as tundra climates, a grouping of some significance with respect to periglacial mass movement processes.

Ecologists and biogeographers have recognised the importance of the tree line in defining the limits of the Arctic and Sub-Arctic climatic-vegetational zones. Blüthgen (1970) identifies a distinctive subarctic zone lying between the arctic and the 'rational' or 'reproductive' forest limits of the coniferous forest belt (figure 1). This limit is where on average seed ripening is possible less frequently than once every five years, so that beyond it large breaks occur between successive generations of trees. The subarctic zone so defined includes the continuous herbaceous tundra at the border of the Arctic, the boggy forest tundra, and the northernmost facies of the mountain birch above the 'rational' tree line.

The subarctic is divided by Blüthgen into pararctic and paraboreal (figure 1), the former comprising herbaceous tundra, with average temperatures in the warmest month 4° to 8°, and in the coldest month less than -8°C. Here permafrost is continuously developed. The

8

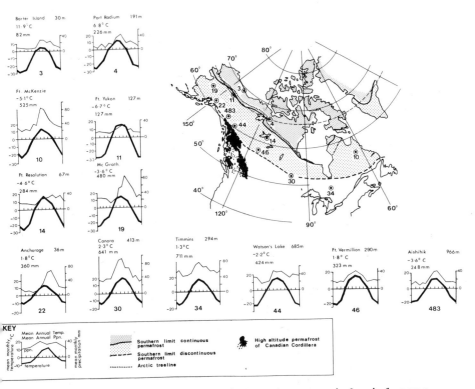

<figure>

Figure 2 Climatic data, North American periglacial zone. Data from Walter and Lieth (1967).

</figure>

paraboreal is subdivided into Oceanic and Continental units. The Oceanic paraboreal includes the subpolar meadows with birch in Alaska, S.W. Greenland, Iceland and Scandinavia, north of about 68°N. Here the warmest month is between 10°C and 12°C, and the coldest 2°C to 0°C. Permafrost is absent, although palsas occur in wetter areas. The continental paraboreal zone is divided into northern and southern units, the northern belt includes the forest tundra of North America and Eurasia, where the warmest month is between 8°C and 10°C and the coldest is below -8°C. Permafrost is generally present. The southern continental paraboreal zone includes open, often boggy forest stands beyond the rational forest limit. The warmest month shows 10°C to 12°C, and the coldest below -8°C. Permafrost is often present.

In addition to these major subdivisions of the subarctic, Blüthgen includes the mountainous vegetation zones, where trees are absent, and vegetation consists of open tundra communities.

The contrast in location of the 'tree line' in figure 2, taken from Brown (1967) and the rational forest limit of Blüthgen (figure 1) is apparent, and reflects the fact

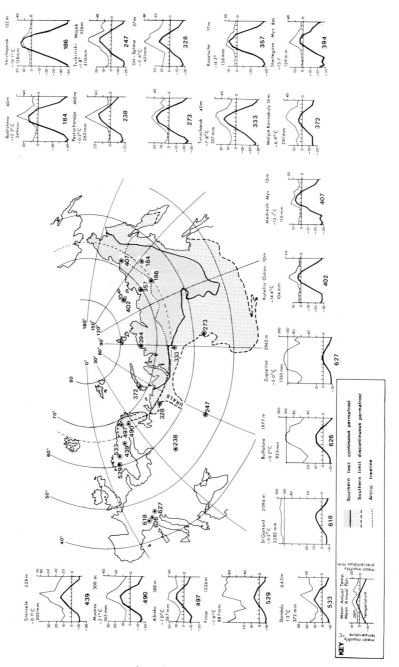

Figure 3   Climatic data, Eurasian periglacial zone.   Data from Walter and Lieth (1967).

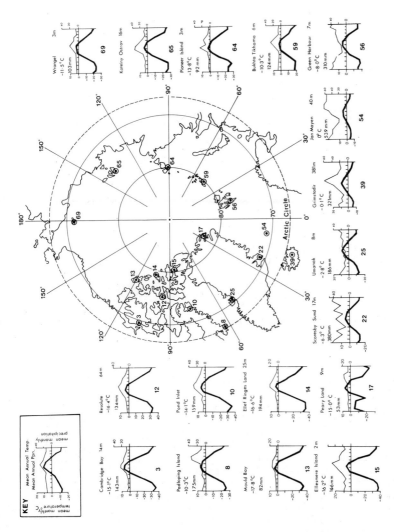

Figure 4 Climatic data, the Arctic. Data from Walter and Lieth (1967).

11

that the northern limit of trees is not a clearly defined
boundary. Hustich (1966) defines three main limits; the
economic forest line, or rational forest limit as defined
by Blüthgen, the physiognomic forest line, which is the
limit of the forest regardless of its reproductive
capacity, and the tree line, which is understood to mean
the absolute polar or altitudinal limit of trees,
regardless of species. Between the outermost limit of
trees (tree line) and the forest line (physiognomic
forest line) is a transitional belt in which isolated
trees and small groves, or larger islands of forest occur.
This transitional belt is mostly called the forest tundra.
Hustich suggests that this forest tundra constitutes the
phytogeographical subarctic zone. He suggests that the
physiognomic forest line is more easily identified in the
landscape than the 'rational' forest line, and therefore
provides a more useful means of defining the boundary
between boreal forest and subarctic forest tundra.
Therefore the subarctic zone as defined by Hustich is less
extensive than that defined by Blüthgen (figure 1).

Simple classifications of climates infact hide
considerable diversity with many intermediate types lying
on the boundaries between recognised climatic zones.
Some indication of this diversity is given with mean
monthly rainfall and temperatures over the Northern
Hemisphere periglacial zones in figures 2, 3 and 4 the
data taken from Walter and Lieth (1960). Walter and Lieth
group cold climates into Arctic, Boreal and Mountain
types, but include many transitional zones on their maps.
The effects of continentality and relief on temperatures
is well illustrated in the North American and Eurasian
maps (figures 2 and 3) and the importance of latitude is
shown clearly in the Arctic map (figure 4) with very short
summer periods of above zero average temperatures, and
severe winter temperatures in the High Arctic Islands.

Vegetation

The tree line is not only climatically significant, it
also appears to define the zone where solifluction is the
main process of mass movement. Brown and Péwé (1973)
observe that active solifluction occurs almost exlusively
above and beyond the forest limits in the North American
Arctic. Péwé (1975) shows that in Alaska solifluction is
common in the central and northern part of the state, and
illustrates the relationship between the incidence of
solifluction and the tree line in central Alaska (figure
5). Similarly, Williams (1961) although accepting that
some solifluction may be possible under tree cover, states
that large scale solifluction, giving terrace forms,
cannot occur where there is heavy tree growth. In his
discussion of the distribution of patterned ground in
Sweden J. Lundqvist (1962) also demonstrates a marked
relationship between the lower limit of sorted and
nonsorted steps (solifluction lobes and terraces) and the
upper limits of the sub-alpine birch zone (figure 6) and

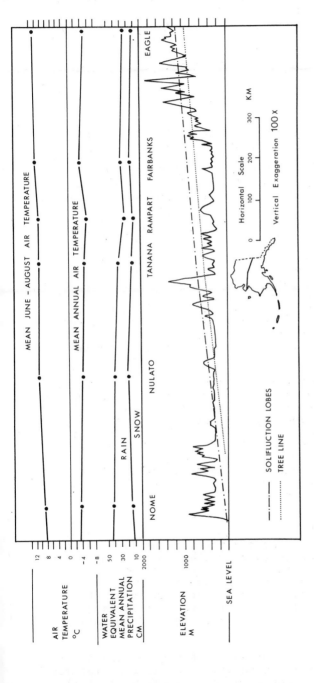

Figure 5   East-west section across central Alaska illustrating change of certain meteorological parameters, elevation of solifluction lobes and the present tree line (Péwé 1975).

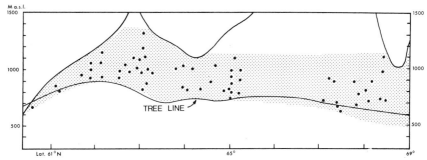

Figure 6   Altitudinal range of solifluction lobes and
            terraces, Sweden (Lundqvist 1962).

Harris (1976) in a detailed study of the Okstindan
Mountains in North Norway, shows a similar correspondence
between the treeline and the lower limit of solifluction
features (figure 7).

A search of the literature tends to confirm that
solifluction is widespread poleward of the arctic tree
line and above the alpine tree line but is noteable for
its absence in the boreal forest zone. In the forested
areas with continental periglacial climates mass movements
are often associated with the thawing of frozen ground
but these failures are generally rapid slides such as
skinflows, bimodal flows and slumps. Such failures leave
a scar on the landscape by disturbing both the ground
surface and the vegetation, and are therefore readily
discernable.

Raup (1951) describes the disturbance of trees in the
boreal forests by mass movements of the soil associated
with thawing of frozen ground. When soil masses move
they tend to tear the root system, killing the trees, so
that it is not uncommon to find 25% of the standing trees
in a forest dead. Examination of the growth rings show
sudden suppression of growth, with subsequent gradual
adjustment. A characteristic of these forests is the
large number of leaning trees which have been tipped over
by the movement of the soils.

Frozen Ground

A common denominator between the various forms of
periglacial mass movements is that they are all associated
with the thawing of frozen ground (McRoberts and
Morgenstern 1974). Such processes as skinflows, bimodal
flows and slumps are restricted to areas underlain by
permafrost. Smaller scale mudflows and solifluction
however occur in both the permafrost zone and in the zone
of deep seasonal freezing.

In their review of permafrost in North America Brown
and Péwé (1973) note that in addition to air temperatures,
vegetation, relief and snow cover are important factors

14

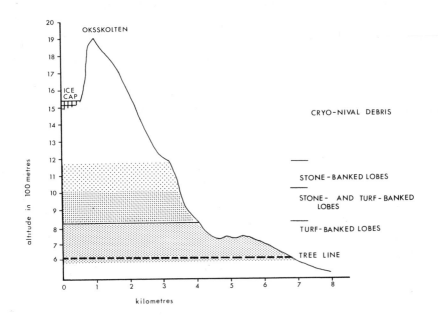

Figure 7   Altitudinal range of solifluction features, northern flanks of Oksskolten, Okstindan, Norway.

influencing particularly the southern limit of discontinuous permafrost. In Canada this limit appears to coincide roughly with the -1°C mean annual isotherm, but in Alaska the southern limit of permafrost extends south of the 0°C isotherm due to the presence of numerous bodies of relict permafrost not in equilibrium with the present climate. The boundary between the discontinuous and continuous permafrost zones is taken by Brown and Péwé to correspond to the -8.5°C mean annual isotherm in Canada, but in Alaska the boundary is again south of this line. The -8.5°C isotherm of mean annual air temperature corresponds with a mean annual ground temperature of -5°C, which according to Soviet studies is the temperature at the depth of zero amplitude at the southern limit of continuous permafrost in Eurasia.

Outside the permafrost zone significant mass movements associated with the thawing of frozen ground require deep annual freezing of the ground. Williams (1961) suggests a minimum depth of frost penetration for solifluction of 70cm. He stresses the importance of summer temperatures in inhibiting frost penetration in winter, and points out that an area with continental extremes of winter cold and summer warmth will suffer shallower frost penetration than might be expected from the winter freezing index (degree Centigrade hours below zero), because of the summer heat in the ground carried over into the following winter. Therefore for a given

Table 3 Climatic data and frost action.

Location	Lat.	Long.	Approx. Frost index value °C hours	Mean Monthly temp. coldest month °C	Mean Ann. air temp. °C	Solif- luction and patt- erned ground	Clim- atic type
Ottawa, Canada	45°28'N	75°38'W	24,000	-11	+5.0	None	Some- what Cont- inental
Calgary, Canada	51°2'N	114°2'W	30,000	-10	+4.0	None	Cont- inental
Trollheimen Norway	62°45'N	9°13'E	25-30,000		-1.0	Solif- luction	Some- what Mari- time
Godthaab, Greenland	64°11'N	51°43'W	25,000	- 8	-0.7	Humm- ocks	Mari- time
S.E. of Snaefell, Iceland	64°40'N	15°25'W	12,000	- 3	+1.0	Solif- luction	Mari- time

(from Williams 1961)

winter frost index value deep freezing of the ground and solifluction are more likely in maritime climates which lack high summer temperatures than in more continental climates (table 3).

The limits of deep frost penetration (greater than 70cm) correspond roughly to a mean annual ground temperature of +4°C, or mean air temperatures one or two degrees lower. Local conditions of vegetation, snow cover and soil thermal properties however produce wide local variations from this trend. Williams also points out that for large scale solifluction to occur thawing must be from the surface downwards rather than from below. The flow of heat from below in spring is closely related to the mean annual ground temperature, and decreases as this approaches 0°C. Williams therefore suggests a mean annual temperature of about +1°C as the rough limit of significant solifluction in both maritime and continental climates.

2. THE SOIL THERMAL REGIME

Introduction

The main characteristic of periglacial slopes as distinct from slopes in non-periglacial areas is the presence of frozen ground for at least part of the year. In high latitudes and at high altitude, the depth of winter freezing generally exceeds the depth of summer thawing and permafrost occurs below the active layer. In lower latitudes summer warmth is greater, and soil frozen in winter may be completely thawed during summer. Under the latter circumstance frozen subsoil persists during the thaw period, but is eliminated once thawing from the surface reaches the base of the frozen layer. The major effect of a frozen subsoil is to impede drainage of the active layer during thaw and hence promote instability on slopes. In permafrost areas it is the depth of thawing in summer which limits the thickness of the active layer within which water-saturated mass movements occur. In non-permafrost areas the depth of winter freezing limits the thickness of the active layer and hence the depth to which such mass movements can take place.

In addition to the depths of soil freezing and thawing, the rates of freezing and thawing are also important. The development of segregation ice during soil freezing may be influenced by the rate of soil freezing (chapter 3), and the generation of excess pore pressures during thaw, with resulting soil instability, may depend to a large extent on the rate of thawing (chapter 4).

The nature of the soil thermal regime is therefore a major factor in periglacial slope stability.

Soil freezing

If it is assumed that all the water present in a soil freezes at a temperature of $0^{\circ}C$, a simple thermal energy diagram illustrates heat losses during freezing (figure 8a). As the unfrozen soil is cooled, heat must be removed at a rate of C_{vu} units per unit volume of soil per degree fall in temperature, where C_{vu} is the volumetric heat capacity of the unfrozen soil. When freezing begins latent heat of fusion must be removed at a rate of L heat units per unit weight of water changed from liquid to solid. The soil temperature remains nearly constant until the pore water is frozen, this period being referred to as the 'zero curtain' in field studies. Following phase change the temperature continues to fall as heat is removed, the rate depending on C_{vf} the volumetric heat capacity of the frozen soil.

This, however, is a simplified picture of the freezing of moist soil. Moisture migration takes place

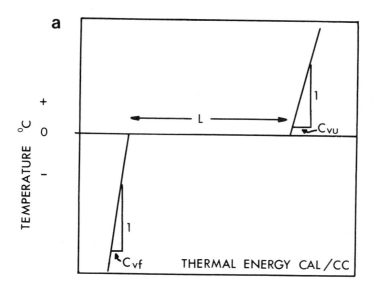

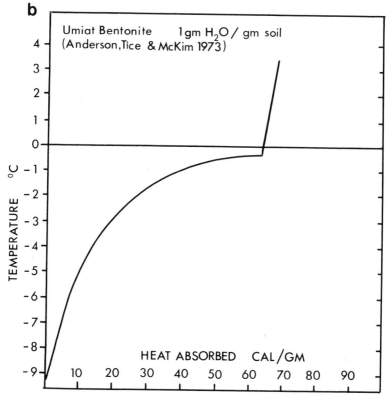

Figure 8 (a) Idealized thermal energy diagram of soil freezing.
          (b) Thermal energy diagram of fine grained soil.

Table 4   Freezing point depressions

	Water content gm/gm soil		
	0.25	0.5	1.0
	Freezing point depression θ, $^{\circ}C$		
Manchester very fine sand	1.22×10^{-18}	1.00×10^{-25}	1.00×10^{-31}
Fairbanks silt	6.37×10^{-3}	7.60×10^{-4}	9.07×10^{-5}
Kaolinite	8.72×10^{-1}	1.27×10^{-1}	1.85×10^{-2}
Suffield silty clay	1.56×10^{-1}	7.37×10^{-1}	1.91×10^{-3}
Hawiian clay	2.91	1.68×10^{-1}	9.70×10^{-3}
Umiat bentonite	18.2	2.41	3.19×10^{-1}

from Anderson, Tice and McKim (1973).

within the cooling soil while it is unfrozen, to complicate its thermal properties. In particular, water may migrate to the freezing plane from the unfrozen soil below, to be incorporated into growing ice lenses. This increases the amount of latent heat which must be removed before advance of the freezing plane can take place.

In addition, not all the soil water freezes and thaws at $0^{\circ}C$. This is due to the effects of surface tension in the films of water wetting soil grains and filling fine capillary pores. The freezing point depression depends on the thickness of the water film. Freezing at $0^{\circ}C$ affects the free gravitational water in the larger pores, but the resulting ice is separated from the mineral grains by a layer of unfrozen water (Anderson et al. 1973). Progressive cooling below $0^{\circ}C$ results in the progressive freezing of the water film so that it becomes thinner and the unfrozen water content of the soil decreases. Latent heat of fusion must therefore continue to be removed even at temperatures below $0^{\circ}C$ while the unfrozen interfacial water freezes (figure 8b).

Free gravitational water may not be present in fine grained soils due to the small size of the pores, and freezing may therefore not be initiated until soil temperatures are reduced below $0^{\circ}C$. Anderson et al. (1973) calculate the freezing point depression for soils ranging from fine sand to clay (table 4). Freezing point depression is greatest for lower moisture contents because of the increased proportion of water present as fine capillary films. For sandy soils freezing point depression is slight, but it becomes much greater for clays, particularly at lower moisture contents.

The thermal gradient within the soil which causes heat flow is generated by soil surface temperatures higher or lower than those of the soil below. Soil surface temperatures are the result of the interaction of such factors as air temperature, solar radiation, terrestrial radiation, evapotranspiration, and the insulating effect of vegetation and snow.

Thermal properties of the soil

The main soil thermal properties affecting the rate of cooling and warming are the heat capacity and the thermal conductivity. Heat capacity controls the rate of change of temperature for a given heat loss or gain, and the thermal conductivity controls the rate of heat loss or gain with a given thermal gradient.

Heat capacity

The heat capacity of soil may be expressed in terms of unit weights (Specific Heat Capacity) or in terms of unit volumes (Volumetric Heat Capacity) and is defined as the heat lost or gained per unit of soil per degree change in temperature. Units of specific heat capacity are calories per degree C per gram, or kilojoules per kilogram per degree K, and of volumetric heat capacity, calories per degree C per cc, or kilojoules per metre^3 per degree K.

The heat capacity of a soil is equal to the sum of the heat capacities of its constituents, mineral material and water in the unfrozen state, and mineral material and ice in the frozen state. Kersten (1952) points out that the specific heat capacity of many rocks is around 0.17 cal/°C /gm (0.71 kJ/kg/°K) while that of water is 1 cal/°C/gm (4.2 kJ/kg/°K). Clearly the moisture content has a major impact on the specific heat capacity of the soil. The specific heat capacity may be found from the percentages by weight of mineral material and water by the simple formula:

$$C_u = \frac{100C_m + C_w W}{100 + W} \qquad (2,1)$$

where C_m = Specific heat capacity of mineral soil

C_w = Specific heat capacity of water, and

W = Moisture content as % dry weight

The volumetric heat capacity of moist soil is given by:

$$C_{vu} = \gamma_d(C_m + C_w W/100) \qquad (2,2)$$

where γ_d = dry density

For frozen soils, if it is assumed that all the soil water is frozen, the heat capacities may similarly be found by substituting the specific heat capacity of ice (approx. 0.5 cal/°C/gm, 2.1 kJ/kg/°K) for that of water in equations 2,1 and 2,2 above.

When taking into account the presence of unfrozen water in soils below 0°C it is convenient to combine the specific heat capacity of the soil/ice/water mixture with the latent heat released by freezing or absorbed by thawing during cooling or warming. The term 'apparent specific heat capacity' is employed by Williams (1964) and Anderson et al. (1973) to describe the heat capacity of a frozen soil plus the latent heat involved in phase changes consequent upon changing its temperature.

The apparent specific heat capacity is given by Anderson et al. (1973) as:

$$C_a = C_m + C_i(W_w - W_u) + C_{wu}W_u + 1/\Delta T_T\int_1^{T_2} L_i(dW_u/dT)\,dT \qquad (2,3)$$

where C_a = apparent specific heat capacity, W_w = total water content W_u = unfrozen water content C_m = specific heat capacity of soil matrix, T = the temperature in °C C_{wu} = specific heat capacity of unfrozen water, $\Delta T = T_1 - T_2, C_i$ = specific heat capacity of ice, and and L_i = latent heat of fusion.

Anderson et al. (1973) also show that the unfrozen water content is well represented by a simple power equation:

$$W_u = \alpha\theta^\beta \qquad (2,4)$$

where θ = temperature in degrees C below zero and, α and β = parameters characteristic of each soil and obtained by regression analysis.

Substituting equation 2,4 in equation 2,3 and intergrating, gives:

$$C_a = C_m + C_iW_w + (C_u - C_i)\alpha\theta^\beta + (\alpha L_i/\Delta\theta)(\theta_1^\beta - \theta_2^\beta) \qquad (2,5)$$

where the limits of integration are taken as: $\theta_1 = [\theta - (\Delta\theta/2)]$ and $\theta_2 = [\theta + (\Delta\theta/2)]$; $\Delta\theta = (\theta_2 - \theta_1)$.

Knowing the values of α and β, and by adopting suitable values for $\Delta\theta$ it is possible to calculate the apparent specific heat capacity of a frozen soil at any temperature and to construct curves of apparent specific heat against temperature. The heat absorbed (+ve) or extracted (-ve) may then be found, with due allowance for sign, by means of the following equation (Anderson et al. 1973):

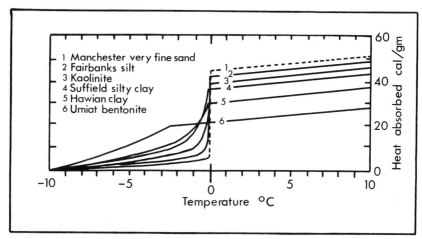

<u>Figure 9</u>   Cumulative heat absorbed in raising frozen soil-
water mixtures containing 0.5 gm H_2O per gm soil
through the melting point (Anderson, Tice and
McKim 1973).

$$Q_{\theta_i \to \theta_j} = \int_{\theta_i}^{\theta_f} C_a d\theta + (C_m + C_w W_w)(\theta_j - \theta_f) \qquad (2,6)$$

where C_w = specific heat of water, and

$\theta_i < \theta_f < \theta_j < 0^0C.$

Using this equation Anderson et al. calculate curves of
cumulative heat absorbed in raising the temperature of six
representative soils with various moisture contents, through
0^0C. The curves for moisture content of 50% dry weight are
redrawn in figure 9. The gradual thawing of ice in the finer
grained soils as the temperature approaches zero is clearly
illustrated, while for sandy soils nearly all the pore water
thaws near to 0^0C. In sandy or silty soils therefore the
assumption that all the soil water freezes at 0^0C may not
introduce significant errors into frost penetration calculations and specific heat and latent heat may therefore be
considered separately. For clay soils however unfrozen
water at temperatures below 0^0C and consequent latent heat
exchange on cooling or warming may make it necessary to
consider apparent specific heat capacities, and to abandon
the concept of a freezing or thawing plane at 0^0C in favour
of a freezing or thawing zone of finite thickness with a
temperature range which may extend well below zero.

<u>Thermal conductivity</u>

A useful basis for discussion is to consider the thermal conductivities of various soil constituents (table 5).

Significant contrasts in thermal conductivity are

Table 5 - Thermal conductivities

		W/m/$^{\circ}$K		cal/sec/cm/$^{\circ}$C
Air		0.024		0.000057
Water		0.602		0.00144
Ice		2.22		0.0053
Shale		1.46		0.0035
Granite	cf.	2.7	cf.	0.0064

between air and water and between water and ice. It is apparent that the relative proportions of the major constituents, mineral grains, water and air is the major factor influencing soil thermal conductivity. The major variable in a given soil is its moisture content, and all investigators report increased thermal conductivity with increased moisture content (e.g. Kersten 1948, Smith 1939), the increase continuing up to the point of saturation (Kersten 1952).

The thermal conductivity of a soil is also directly related to its dry density since increases in dry density result in closer contacts between mineral grains and a reduction in pore space.

Temperature is a third factor influencing the thermal conductivity of soils when soil freezing takes place. The thermal conductivity of ice is 2.22 W/m/$^{\circ}$K (0.005 cal/cm/sec/$^{\circ}$C) while that of water is 0.602 W/m/$^{\circ}$K (0.0014 cal/cm/sec/$^{\circ}$C) so that when soil freezing occurs the thermal conductivity increases by an amount dependant on the moisture content. Penner (1970) investigated the thermal conductivity of fine grained soils in which the water does not all freeze at 0°C, but where ice content increases at the expense of water content as the temperature falls below 0°C. He shows that at -10°C Leda clay contains approximately 10% unfrozen water, while Sudbury loam contains about 5%. Figure 10 shows thermal conductivities of initially saturated Sudbury loam and Leda clay. It is apparent that the thermal conductivities are temperature dependent at temperatures below 0°C, particularly at temperatures close to zero. However, Williams and Nickling (1972) suggest that changes in thermal conductivity during freezing are relatively small, and for some practical considerations will be of relatively little significance.

The effects of variation in moisture content, dry density, and phase composition differ for soils of different textures. Laboratory studies (Baver 1956) show

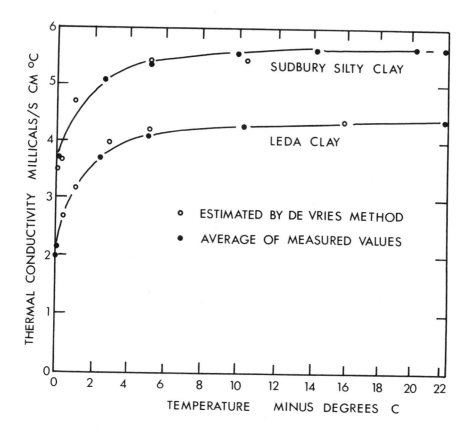

<u>Figure 10</u> Estimated and measured thermal conductivity
for partially frozen soils (Penner 1970).

that at a given dry density coarse grained soils show
higher conductivity values than finer grained soils,
although Smith (1939) states that soils containing high
organic contents show lowest conductivity values.
Generally, decreased grain size is associated with
increased void ratio, increased water/air content and
consequently decreased thermal conductivity.

Kersten (1948, 1949, 1952) from laboratory
measurements at the University of Minnesota produced
charts from which thermal conductivity estimates may be
made from knowledge of texture, moisture content,
temperature and dry density. Soils are subdivided into
silt and clay soils (fine textured) and sandy soils
(coarse textured). The charts shown opposite (figure 11)
have been converted from Kersten's original Imperial
Units.

These charts can provide only a rough guide to
conductivities because of generalizations with regard to
texture, soil structure, mineral composition and organic

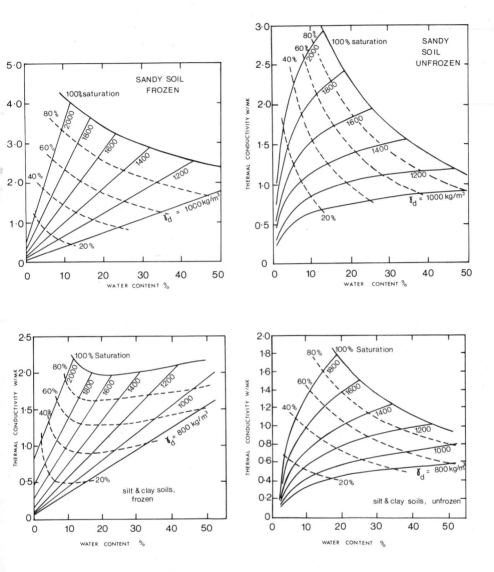

Figure 11　Average thermal conductivities of moist soils
derived from Kersten's basic work (Harlan and
Nixon 1978).

content.　However, Nixon and McRoberts (1973) show that
for many problems of geotechnical and geomorphological
interest, methods of calculation of frost penetration are
relatively insensitive to slight errors in the value of
thermal conductivity used.　Where data on the thermal
conductivity of a particular soil are not available
therefore, thermal calculations utilizing conductivities
estimated from Kersten's charts may still be successfully

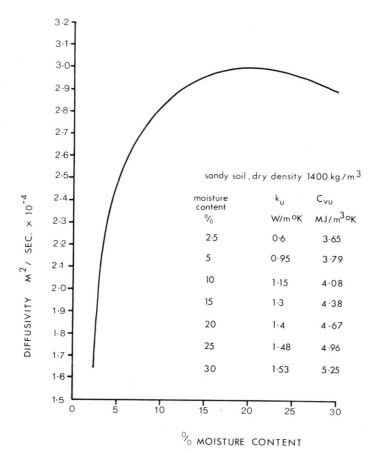

The table within the figure:

sandy soil, dry density 1400 kg/m³

moisture content %	k_u W/m°K	C_{vu} MJ/m³°K
2.5	0.6	3.65
5	0.95	3.79
10	1.15	4.08
15	1.3	4.38
20	1.4	4.67
25	1.48	4.96
30	1.53	5.25

Y-axis: DIFFUSIVITY M²/ SEC. × 10⁻⁴

X-axis: % MOISTURE CONTENT

Figure 12   Thermal diffusivity of a hypothetical sandy soil, k_u values from figure 11, C_{vu} values calculated.

applied.

Thermal diffusivity

The rate of heating or cooling of a given soil depends on its thermal conductivity and its heat capacity. The thermal diffusivity combines these two parameters, where:

$$a = k/C_v \qquad (2,7)$$

and a = thermal diffusivity, k = thermal conductivity, and C_v = volumetric heat capacity. The units of thermal diffusivity are cm²/second or m² per second.

Consideration of the thermal diffusivity illustrates the importance of moisture content in the rate of soil cooling under a given thermal gradient (figure 12). As the moisture content increases the thermal conductivity increases, but so does the volumetric heat capacity.

Initially the rate of increase in k is greater than the rate of increase in C_{vu} so that diffusivity increases and the soil cools more rapidly. However, at higher moisture contents the increases in heat capacity become greater so despite slight increases in thermal conductivity, the thermal diffusivity of the soil decreases, and the rate of cooling slows down. At some intermediate moisture content therefore the soil will have maximum diffusivity, at higher moisture contents the diffusivity is less due to high volumetric heat capacity, at lower moisture contents the diffusivity is less because of low thermal conductivity.

If, however, the apparent volumetric heat capacity of a soil is considered the thermal diffusivity will be much lower during freezing than in the frozen or unfrozen state where no phase change occurs with change in temperature. The apparent volumetric heat capacity includes both the heat capacity of the soil constituents and the latent heat released on phase change. Williams (1964) and Williams and Nickling (1972) provide diagrams showing apparent volumetric heat capacities for Niagra silt and Leda clay (figure 13). The apparent heat capacities during freezing are clearly far greater than those of the unfrozen and frozen soils, and are temperature dependant, and depend on whether the samples are warming or cooling. Thus the thermal diffusivities are greatly reduced during the freezing process, and the penetration of freezing through the soil is much slower than the penetration of a wave of heating or cooling where no phase change takes place.

It should be noted that for soils with high moisture contents the diffusivity of the frozen soil is appreciably higher than that of the unfrozen soil because the thermal conductivity of ice is nearly four times that of water and the volumetric heat capacity approximately half that of water. Kersten (1952) shows that at moisture contents of 15% for example, the diffusivity of a frozen soil may be 50% higher than that of the unfrozen soil.

Influence of moisture migration

Moisture migration in soils tends to take place in response to a thermal gradient. Bouyoucos (1915) studied moisture migration and suggested that viscosity and surface tension of soil water decreases as the temperature rises, the viscosity decreasing much more rapidly than the surface tension. It was believed that as surface tension increased during cooling, water was drawn towards areas of higher tension (cooler areas) away from areas of lower tension (warmer areas). Such moisture migration must complicate the thermal properties of a cooling soil. Williams and Nickling (1972) also point out the importance of moisture migration to the frost line during soil freezing. Such migration, associated with the development of ice lenses in the soil, increases the latent heat

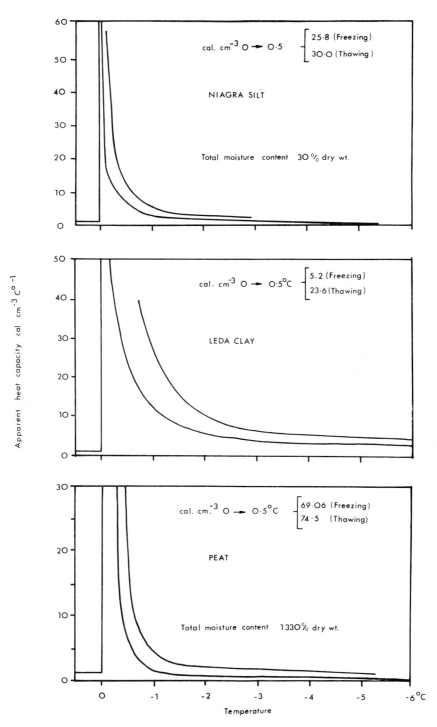

Figure 13   Apparent heat capacities during freezing
             (Williams and Nickling 1972).

28

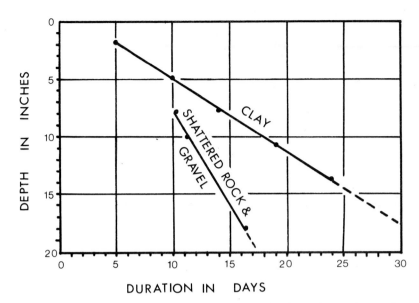

Figure 14   Duration of zero curtain during freezing of
             soils at Resolute Bay, N.W.T., Canada, 1955
             (Cook 1955).

released at the freezing plane, and the migrating water
tends to transport sensible heat to the freezing plane.
Cook (1955) illustrates well the contrasting rates of
soil freezing in coarse non frost heaving gravel and frost
susceptible clay. In the autumn freeze of 1955 the depth
of freezing on the 12th September at Resolute Bay, North
West Territories, was 5 in (12.7 cm) in the clay soil, but
12 in (30.48 cm) in the gravel. Growth of ice lenses in
the soil also prolonged the period of freezing, or 'zero
curtain! effect at each monitored depth in the clay soil,
as shown in figure 14.

External factors affecting soil freezing and thawing

Air temperatures

Winter soil freezing results from heat loss to the
atmosphere when air temperatures are below freezing. The
thermal gradient in the soil depends upon soil surface
temperatures, which during freezing are commonly higher
than air temperatures by as much as 4°C (Williams and
Nickling 1972). This is due to the insulating effect of
vegetation cover, but where snow cover is also present
the contrast between air temperatures and soil surface
temperatures may be even greater. In January 1970 in the
Okstindan Mountains of Norway, Harris (1974) observed a
soil surface temperature of -2.25° beneath a snow cover
of 72 cm, when air temperatures were as low as -28°C.

It has been shown however that the depth of winter
soil freezing is proportional to the below-zero air

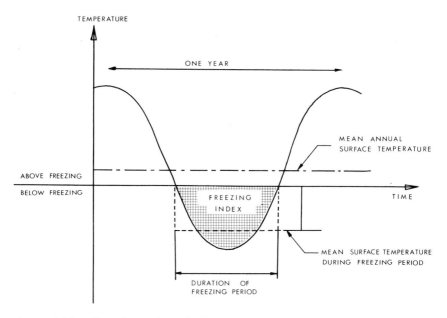

Figure 15   The freezing index.

temperatures and the length of time for which they persist
(Beskow 1947, Aldrich 1956, Kersten 1959). The product
of the mean sub-zero air temperature and the time for
which freezing temperatures persist is known in the
Scandinavian literature as the 'Cold Quantity' and is
expressed in degree hours below zero centigrade. American
literature has in the past expressed a similar 'Freezing
Index' in degree days below 32° Farenheight, but more
recent literature has tended to use the term freezing
index with units of degree days or degree hours below zero
centigrade.   The freezing index is illustrated in figure
15.

Kersten (1959) investigated the penetration of
freezing below highway pavements which were kept clear of
snow. He assumed that the surface temperatures would be
3°F (1.6°C) warmer than the air temperatures on average,
and therefore related frost penetration to the number of
degree days below 29°F(-1.6°C) rather than 32°F. For a
series of test sites he found a close relationship between
the depth of frost penetration and the square root of the
number of degree days below 29°F (figure 16).

In a study of soil temperatures in Northfork,
California (37°N, 840 m a.s.l.) Anderson (1947) related
soil freezing and thawing to various meteorological
parameters. Sites were monitored where the ground was
kept free of vegetation and where vegetation consisted
of grass. By means of regression analysis Anderson
identified minimum air temperature, hours of freezing,
hour-degrees of freezing air temperatures and cloud cover

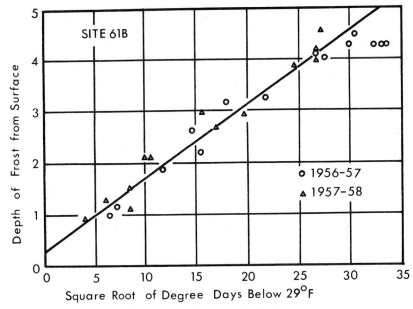

Test Series III; vicinity Mpls. and St. Paul

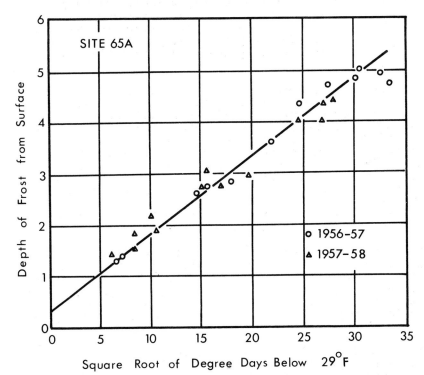

Figure 16   The relationship between the depth of frost below highway pavements and the square root of degree days air temperature below 29°F (Kersten 1959).

31

as significant factors affecting the depth of soil
freezing. The hour-degrees of freezing gave the best
estimate of frost depth in both monitored sites, and
comparison of the regression coefficients for the bare
soil and the grass covered sites showed that an equal
increase in the hour-degree of freezing caused three times
as great an increase in the depth of freezing in the bare
soil as in the grass covered soil.

Smith and Tvede (1977) describe computer simulations
of frost penetration beneath highways utilizing the model
developed by Outcalt (1972). The inputs of weather data
used to estimate surface temperatures are solar radiation,
cloud cover and type, air temperature, wind speed and
atmospheric pressure. The albedo and the aerodynamic
roughness of the highway surface are also considered.

Soil thawing takes place when the soil surface
temperature rises above 0°C. Soil surface temperatures
depend partly on air temperatures, but to a greater
extent on the absorbtion of radiation, both short wave
length solar radiation, and long wave back-radiation from
the atmosphere (Aldrich 1956, Gold et al. 1972). Factors
influencing the soil surface temperature therefore
include the albedo, which varies considerably with colour
and roughness, and vegetation cover, which intercepts
radiation and utilises the heat energy in transpiration.
A thawing index based only on air temperature is therefore
likely to be less useful in predicting soil thawing rates
than a corresponding freezing index for predicting soil
freezing rates.

Anderson (1947) in his study of soil temperatures referred
to above, found solar radiation, the hour-degrees of
thawing air temperatures, cloud cover, and the soil
temperature at 12 in (30.48 cm) depth to be significantly
related to the rate of soil thawing. The solar radiation
factor gave the highest correlation coefficient with depth
of thaw. Partial regression coefficients in a multiple
regression of thawing depth against solar radiation and
hour degrees above zero air temperatures also indicated
that the solar radiation was more highly associated with
soil thawing than was heating of the soil by conductive
and convective heat transfer from the air.

Chambers (1966) showed that a particularly rapid rise
in the temperature of the active layer on Signy Island in
November and December 1963 was the result of strong solar
radiation affecting snow-free surfaces. He also suggested
that the percolation of meltwater downwards through the
active layer may lead to accelerated thawing at greater
depths.

It is apparent that no simple relationship exists
between air temperature and soil surface temperature.
Calculations of the depths and rates of soil freezing and
thawing should therefore be based on freezing and thawing

Table 6   Thermal conductivity of snow (Crawford 1951)

Month	Nov	Dec	Jan	Feb	Mch	Apl
Average Density (gm/cc)	0.139	0.182	0.193	0.189	0.233	0.279
Thermal Conductivity (cal/cm/^{0}C/sec)	2.9×10^{-4}	3.3×10^{-4}	3.7×10^{-4}	3.5×10^{-4}	4.5×10^{-4}	5.7×10^{-4}

indices derived from soil surface temperatures rather than air temperatures.

Snow cover

A covering of snow has a considerable effect on the thermal regime of the ground because of the low thermal conductivity of snow.   Thermal conductivity depends largely on snow density and Crawford (1951) provides the above table of thermal conductivities and density.

Beskow (1947) estimates that the thermal conductivity of snow is of the order of ten times lower than that of frozen soil, and Gold (1957) quotes values of 3.2×10^{-4} cal/cm/^{0}C/sec for snow of density 0.2 gm/cc and 2.6×10^{-3} cal/cm/^{0}C/sec for Leda clay respectively.   The influence of snow cover on soil temperatures is illustrated diagrammatically by Beskow (1947) (figure 17).   Assuming the thermal conductivity of snow to be ten times lower than that of frozen soil he considers the effect of a layer of snow to be equivalent to an increase in the thickness of frozen soil through which heat must be conducted of ten times the thickness of the snow layer, since:

$$S_f = S_s \cdot \frac{k_f}{k_s} \qquad (2,8)$$

where S_f = thickness of frozen soil, S_s = thickness of snow and k_f and k_s are the thermal conductivities of frozen soil and snow respectively.

This equation however ignores the heat capacities of frozen and unfrozen soil, considering only the latent heat released during freezing, and this results in a slight exaggeration of the influence of snow cover.   However, for moderate snow cover over wet silty or sandy soils figure 17 is a fairly accurate representation of the influence of snow.   It should be noted that in figure 17

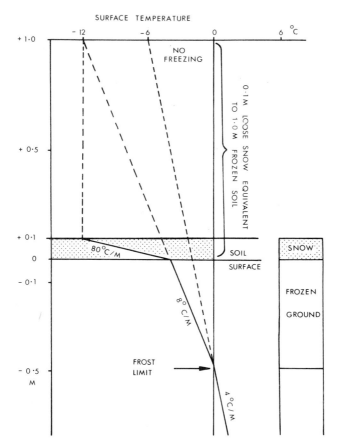

Figure 17   Influence of snow cover on soil temperatures
            (Beskow 1947).

soil freezing will not take place if the snow surface
temperature is higher than -6.0°C because with the
insulating snow cover the rate of heat flux from the soil
is then less than the rate of heat conducted from below.
Such a situation was recorded by Atkinson and Bay (1940)
who found that in two out of three cases where snow depth
was 25 cm or more the frost depth decreased despite sub-
zero air temperatures while where it was less than 25 cm
frost depth increased correspondingly. They concluded
that the depth of frost penetration was inversely
proportional to the depth of snow. Similarly, Gold (1957)
found from a study in the Ottawa area that the rate of
heat flow from the ground was inversely related to snow
depth.

    Gold et al. (1972) present calculations carried out
with a one-dimensional finite difference computer program
to illustrate the effect of surface conditions on the
ground thermal regime. It was assumed that the air

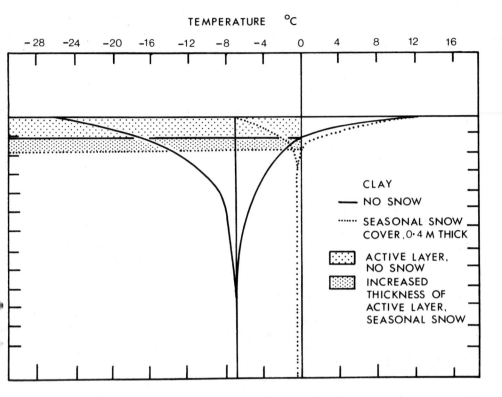

Figure 18   Mean annual ground temperature and envelope of
annual temperature change for clay, with 0.4 m
snow cover, and with no snow cover, showing
the contrast in thickness of the active layer
(Gold, Johnston, Slusarchuk and Goodrich 1972).

temperature varied sinusoidally and had an annual mean of
-7.7°C with an amplitude of 21.3°C.   It was also assumed
that the soil surface temperatures were equal to the air
temperatures.   Snow cover was assumed to begin on 29th
September with the depth increasing linearly until 1st
February when it remained constant until the onset of the
melt period.   These conditions were considered
representative of Inuvik, North West Territories.   The
influence of a snow cover of maximum thickness of 0.4 m
over a clay soil is illustrated in figure 18.   The average
annual ground surface temperature is raised by 6.5°C as a
result of this snow cover, and the thickness of the active
layer is increased.

The importance of snow cover in insulating the ground

Table 7 Soil temperature data, Okstindan, Norway, winter
of 1969-1970.

	Date	Snow Depth cm	Soil temperature, 5 cm depth ^{0}C	Air temperature 0^0C
Site A	25.1.70	72	-2.25	-28
Site B	25.1.70	0-5	-12.6	-28
Site A	4.4.70	85	-2.8	-12.4
Site B	4.4.70	0-5	-15.3	-12.4

from low winter air temperatures is emphasised by Harris
(1974) in his study of soil temperatures in the Okstindan
Mountains of northern Norway. The study area lies above
the tree line at an altitude of around 710 m, where
vegetation consists of heath and sedge communities. The
mean annual temperature is estimated to be around -2.5^{0}C,
but no permafrost is present in the study area.
Solifluction during the spring thaw has produced well
developed turf-banked lobes. During the winter of 1969-70
two sites were monitored with contrasting snow conditions
(table 7). In one site snow drifted to depths of between
75 cm and 1 m but deflation maintained virtually snow-free
conditions at the other site.

The contrast in soil temperatures under the
contrasting snow covers is well illustrated, with soil
temperatures at the 5 cm depth in the snow accumulation
site being some 10^{0}C higher in January and 12.5^{0}C higher
in April than in the site with only limited snow cover.
Harris also notes that during the autumn of 1970 snow
accumulation in Site A reduced the rate of penetration
of frost in the soil to almost half that in the snow free
location Site B, the rates of penetration being
respectively 1.23 cm per day and 2.33 cm per day.

In the Russian Taiga zone Izotov (1967) observed
that snow depths are generally greater in the forested
areas than in open cut-over areas, and consequently
freezing depths are less. Over the period 1961-66
freezing penetrated to 30-34 cm in the monitored forest
sites, compared with 45-60 cm in the open cut-over sites.
Thawing from the surface did not begin until snow had
cleared, although thawing of the frozen soil from below
by geothermal heat began earlier in the spring.

Finally, the importance of snow cover in preventing
permafrost development in the area of Schefferville which
lies in the discontinuous permafrost zone of the
Labradore-Nouveau Quebec Peninsula, Canada has been
examined in detail by Nicholson and Granberg (1973). They
relate mean annual ground temperatures to snow depths
within circular areas of different radii above the ground

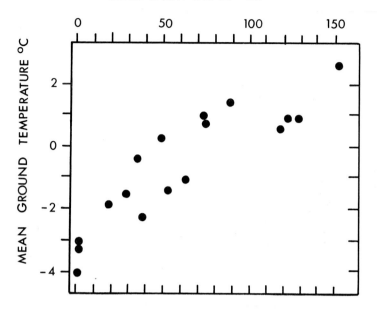

Figure 19   Plot of mean snow depth against mean annual
soil temperature at 1.5 m depth, Schefferville
(Nicholson and Granberg 1973).

temperature recording sites. Figure 19 shows a plot of
mean snow depth against mean annual ground temperatures
at 1.5 m depth, based on 17 recording sites. It is
apparent that the relationship is nearly linear, although
the authors suggest that an exponential relationship
might be expected. However, they point out that on sites
with extremely shallow snow the severe conditions in
winter prevent vegetation being established, and the
resulting bare ground causes higher summer ground
temperatures. Therefore, the very low winter temperatures
are partially compensated by higher summer temperatures.
At the 1.5 m depth the regression equation is $T = 0.0346S$
$- 2.67$, where T = mean ground temperature and S = snow
depth (cm). When T is equal to $0°C$, S is given as 77.2
cm, and regression equations for ground temperature at
different depths against snow thickness show similar
values of S for $T = 0°C$. The authors therefore conclude
that approximately 75 cm of snow is sufficient to prevent
permafrost formation in the Schefferville area.

Since snow depths in winter not only affect soil
freezing rates and depths but also greatly influence soil
moisture conditions during the spring thaw, it must be
concluded that the distribution of snow is a major factor
in the distribution and rates of mass movements on slopes
in periglacial areas.

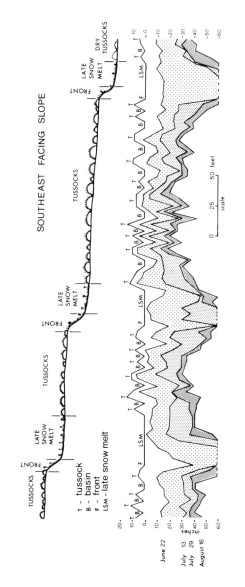

Figure 20 Active layer thickness on a southeast facing slope in the Ruby Range, Yukon, beneath turf-banked solifluction lobes, and its relationship to vegetation (Price 1971)

Vegetation

Vegetation, like snow cover, provides an insulating layer between atmosphere and ground surface, and thereby modifies the heat exchanges between them. Vegetation shields the ground from insolation, by shading; it reduces heat conduction into the ground by reducing air circulation; sensible heat transfers by rain are reduced by interception; and through evaporation and transpiration the vegetation absorbs heat which would otherwise be utilized in warming the soil. Benninghoff (1952) shows that in permafrost areas vegetation plays an important role in modifying heat penetration into the ground in summer, and hence affects the thickness of the active layer. Moss in particular is a good insulator when dry and evaporator when wet, reducing summer thawing of frozen ground and Brown (1971) states that in Arctic Canada the active layer is thinnest in moss covered sites and thickest in areas supporting sedge and other types of vegetation. Pihlainen (1962) describes increases in the thickness of the active layer at Inuvik N.W.T. from 2 feet (approx. 60 cm) in undisturbed moss covered areas, to approximately 5 feet (1.5 m) in areas stripped of moss and underlain by fine-grained soils, and to 8 feet (2.44 m) in areas stripped of moss and underlain by coarse grained soils.

The significance of vegetation to the thermal regime of slopes affected by solifluction was investigated by Price (1971) in the Ruby Range, Yukon Territory, Canada. The thickness of the active layer was found to be closely related to vegetation cover. This relationship is best displayed on south-east facing slopes (figure 20) where the active layer is much thicker below lobe fronts than below tread surfaces. Lobe front vegetation consists of low herbaceous species 5.1 to 10.2 cm high whereas tread surfaces are covered by mossy tussocks, up to 61 cm thick. The greater depth of the active layer below the lobe fronts leads to better drainage of the soil and therefore inhibits moss growth, but beneath the tread surfaces the thin active layer remains poorly drained and mosses are encouraged to grow. At a depth of 2.54 cm in the soil at the lobe fronts, temperatures were on average 7.5°C higher than beneath the tussock community on sunny days, and 1.4°C higher on other days. In addition to the influence of vegetation, the greater inclination of the sun at the steeper lobe fronts also contributed to these contrasting temperatures.

Because of the large variability in vegetation type and thickness, and its observed influence on soil surface temperatures it is rarely sufficient to consider air temperatures as being equal to ground surface temperatures in assessing rates and depths of penetration of freezing and thawing. It is more satisfactory to monitor ground surface temperatures and use these to estimate thermal gradients in the soil.

Calculation of freezing and thawing depths

In discussing the theoretical and practical aspects of the thermal properties of soils Van Rooyen and Winterkorn (1957) pointed out the complexity of heat transfers in natural soils. They state that a 'theory covering all phenomena would be too cumbersome to handle while one simple enough to handle mathematically could not be expressive of the events actually taking place' (p. 144). However, simple methods of estimating depth of frost penetration and thawing have been developed, and more recently the use of computers has made possible more realistic modelling of soil freezing and thawing.

The simplest formula for calculating the depth of frost penetration or the depth of thawing is the Stephan Formula (Stephan 1890), given as:

$$X_f = \sqrt{\frac{2k_f F}{L}} \qquad (2,9)$$

and

$$X_u = \sqrt{\frac{2k_u U}{L}} \qquad (2,10)$$

where X_f and X_u are respectively the depth of freezing and the depth of thawing, k_f and k_u are the coefficients of thermal conductivity of frozen and thawed soil respectively, L is the volumetric latent heat of the soil, F is the freezing index (mean soil surface temperature below freezing multiplied by the duration of sub zero temperatures) and U is the thawing index (mean surface temperature above freezing multiplied by the duration of above zero temperatures).

This is a simplified solution of the Neuman equation (Harlan and Nixon 1978) assuming that the thermal gradient below the 0°C isotherm is zero, and assuming that all the heat supplied to or removed from the soil is latent heat. The latter assumption is based on the fact that the latent heat of moist soil is generally much greater than its heat capacity. In dry mineral soil however, the heat capacity may become relatively important during cooling below the freezing point.

The Stephan equation assumes a sudden 'step' decrease in temperature, defined by the mean sub-zero temperature.

The Stephan equation has been modified to take account of the heat capacity of the soil, and Beskow

40

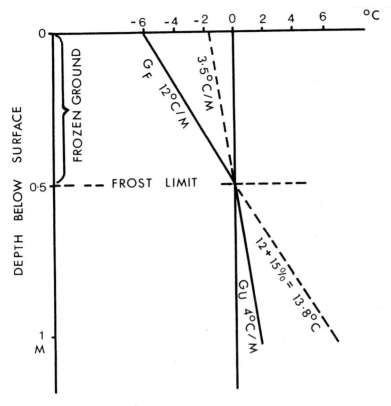

SURFACE TEMPERATURE

Figure 21   The influence of heat conduction from below the freezing plane on the depth of frost penetration (Beskow 1947).

(1947) presents such a formula:

$$X_f = \sqrt{\frac{2k_fF}{Q}}$$

(2,11)

where Q is the 'frost storing capacity of the soil' and is defined as:

$$Q = L_{vi}\,W_v + \frac{t}{2}\,(C_{vi}\,W_v + C_{vm}M)$$

(2,12)

where L_{vi} is the latent heat of ice per unit volume, W_v is the volumetric moisture content, t is the average surface temperature, C_{vi} is the volumetric heat capacity of ice, C_{vm} is the volumetric heat capacity of mineral material and M is the volume of mineral material per unit volume of soil.

This equation may be rewritten as:

$$Q = \frac{L_i W \gamma}{100} + \frac{t}{2} \frac{(C_i W \gamma + C_m \gamma)}{100} \qquad (2,13)$$

where L_i is the latent heat of ice, W is the moisture content as percent dry weight, γ is the density of the dry soil and C_i and C_m are the specific heat capacities of ice and mineral materials respectively. A similar modification of the Stephan equation was developed by Berggren (1943).

Beskow (1947) also illustrated the effect of conduction from below the freezing plane on the depth of frost penetration (figure 21). Assuming a ratio of thermal conductivity of frozen and thawed soil equal to 1.15, Beskow shows that a gradient of 13.8°C/m below the freezing plane would stop frost penetration when the gradient in the frozen soil was 12°C/m, while a gradient of 3.5°C/m in the frozen soil would not cause further frost penetration if the gradient below the freezing plane were 4°C/m. Beskow quotes values of thermal gradients below the freezing plane for southern and central Scandinavia of 7°-10°C/m in December, falling to 2°-3°C/m in the spring.

Summer heat stored in the soil was also considered by Carlson (1952) who included the number of degrees by which the mean annual temperature of the soil surface exceeded the freezing point in a modified Stephan formula, as shown below:

$$X_f = \sqrt{\frac{2k_f F}{L + C_{vu} t_o + C_{vf} t/2}} \qquad (2,14)$$

where t_o is the number of degrees C by which the mean annual soil surface temperature exceeds zero, and t is the mean sub-zero soil surface temperature during the freezing period. The formula has been rewritten to accommodate metric units rather than the Imperial units used in Carlson's original formula. In this version of the modified Stephan equation the amount of heat lost when each unit of soil from the surface to the maximum depth of frost penetration is cooled from the mean annual soil surface temperature down to 1/2 the average sub-zero soil surface temperature during the freezing period is set up for conduction through the average depth of frost.

Where the freezing soil is not homogeneous, but consists of layers with differing thermal properties, Beskow and Carlson adopt similar techniques for predicting frost penetration. Beskow (1947) defines a 'freezing resistance' for each layer as the degree-hours below freezing at the soil surface required to freeze that layer.

42

The freezing resistance $\Omega = t \times T$ where t is the mean surface subzero temperature in degrees Centigrade and T is the time in hours.

and,
$$\Omega_1 = \frac{QX_1{}^2}{2k_1} \qquad (2,15)$$

where X_1 is the thickness of layer 1 and k_1 is its thermal conductivity. Carlson and Kersten (1953) define the term $\frac{X_1}{k_1}$ as the 'thermal resistance' (R) of the layer, being the ratio of the thickness of the layer to its thermal conductivity so that:

$$\Omega_1 = \frac{Q_1 X_1 R}{2} \qquad (2,16)$$

and
$$\Omega_2 = \frac{Q_2 X_2 (R_1 + R_2)}{2} \qquad (2,17)$$

and
$$\Omega_n = \frac{Q_n X_n (R + R_n)}{2} \qquad (2,18)$$

Knowing the total freezing index for any season, or partial season, the depth of freezing may be determined by selecting those layers whose sum of freezing resistance equals the total freezing index for that period. Similar equations may be used for soil thawing if the thawing index U is substituted for freezing index F, and thermal capacities and conductivities for thawed rather than frozen soils are utilized in calculating Q.

The depth of soil freezing is particularly significant in non-permafrost areas where the depth of winter freezing limits the depth to which impedence of drainage by ground ice during the spring thaw can occur. Saturated slope failure therefore, is only likely to take place above this depth. The rate of frost penetration also affects the thickness of ice lenses developed in frost-susceptible soils. The rate of thawing is also highly significant in studies of periglacial slope stability, since the rate of release of excess pore water from its frozen state largely controls the development of pore pressures during thaw consolidation.

Thawing rates have been considered in detail in an important paper by Nixon and McRoberts (1973). They present in some detail Neuman's solution for the penetration of the thaw front and associated temperature fields. They show that the depth of thaw may be given as:

$$X_t = \alpha \sqrt{T} \qquad (2,19)$$

where X_t is the depth of thaw and T is the time, and:

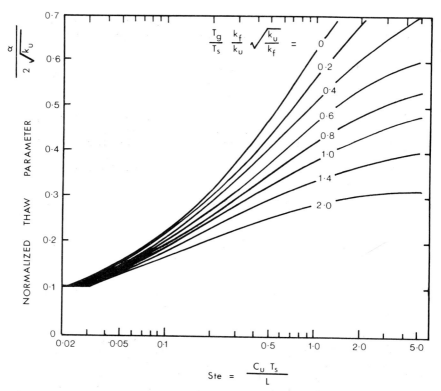

$$\frac{\alpha}{2\sqrt{k_u}}$$ NORMALIZED THAW PARAMETER

$$\frac{T_g}{T_s} \frac{k_f}{k_u} \sqrt{\frac{k_u}{k_f}} = 0$$

0·2
0·4
0·6
0·8
1·0
1·4
2·0

$$Ste = \frac{C_u T_s}{L}$$

<u>Figure 22</u>   Graphical solution of the Neuman equation
(Nixon and McRoberts 1973).

$\alpha = f(k_u \ k_f \ C_{vu} \ C_{vf} \ t_g \ t_s \ L)$ where t_g = ground temperature
and t_s = mean surface temperature.

Nixon and McRoberts provide a graphical solution of
the Neuman equation (figure 22), assuming k_u/k_f = 0.7, a
reasonable assumption for most soil conditions. From
known soil parameters and thermal conditions it is
possible to extract the α value with a high degree of
accuracy from this chart. On a semi-empirical basis, if
the temperatures below the thaw front are close to the
melting point α may also be predicted from the equation:

$$\frac{\alpha}{2\sqrt{k_u}} = \sqrt{\frac{Ste}{2} . (1 - \frac{Ste}{8})} \qquad (2,20)$$

where Ste is the Stephan Number and equals $\dfrac{C_{vu}t_s}{L}$ (2,21)

where C_{vu} is the unfrozen volumetric heat capacity , t_s
is the mean surface temperature, and L is the latent heat
of the soil.

Nixon and McRoberts consider the effect of simplifying

44

assumptions made in calculating thawing rates. They show that for many problems of geotechnical interest neither the absolute magnitude, nor the temperature dependance of the thermal conductivity of frozen fine grained soils has any significant influence on calculated rates of thaw. Secondly, the temperature dependance of the unfrozen moisture content at temperatures between 0^0C and t_g and its effect on the volumetric heat capacity of frozen soils is completely insignificant with regards to prediction of the rate of thaw. Finally, the solution obtained from the Neuman equation by ignoring temperature-dependant latent heat effects due to unfrozen water present in the soil below 0^0C is satisfactory, and introduces insignificant errors in the calculated rates of thaw. However, allowance should be made in the calculation of the latent heat term for the unfrozen water content of the frozen soil at the average temperature below the thaw plane, where:

$$L = \gamma w \ (1 - w_u) \ L_i \qquad (2,22)$$

where L = latent heat of soil per unit volume, γ = dry density of the soil, w = moisture content, gm/gm dry soil, L_i = latent heat ice = 79.6 cal/gm and w_u is the unfrozen water content.

McRoberts (1975) compared published data on the thawing of soils in the field with predicted rates of thaw using the Neuman solution:

$$X_t = \alpha\sqrt{T}$$

Case histories of thaw were analysed by plotting the depth of thaw against the square root of time. Fifteen records were examined and for most there was a marked linear relationship between depth of thawing and the square root of time, with a proportionally constant equivalent to α. McRoberts utilized the Stephan equation to calculate the magnitude of α from the temperatures and thermal properties reported in the fifteen case studies (figure 23). Despite the relatively crude assumptions made in the calculation, and the uncertainties with regard thermal conditions in some of the field studies, the agreement between the measured and calculated values of the thaw parameter was quite good. The overall range in computed values of α was small, ranging from 0.015 to 0.095 cm/sec$^{1/2}$ and this encompassed all the measured values. This tends to support the suggestion made by Nixon and McRoberts (1973) that the rate of penetration of thawing depends largely on the soil surface temperatures and soil moisture content, but is relatively insensitive to other variables.

The calculations described above, based on the Neuman solution for freezing and thawing of soils assume a step-wise fall or rise in temperature at the soil surface in order to predict rates of freezing and thawing. A more realistic approach is to use a time-dependent soil

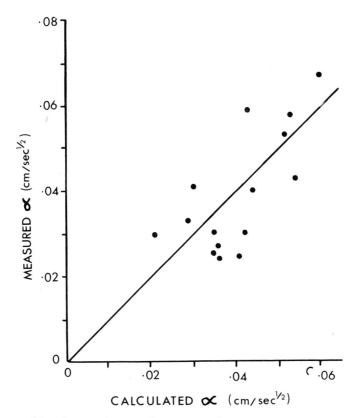

Figure 23 Comparison of measured α with α calculated by Stephan solution (McRoberts 1973).

surface temperature change. Lock et al. (1969) have provided approximate analytical solutions for power law and sinusoidal temperature variations. For the case of the sinusoidal variation in soil surface temperature, the surface temperature t_s is given by:

$$\frac{t_s}{t_{max}} = \sin\left(\frac{T}{T_c}\right) \qquad (2,23)$$

where t_{max} is the peak surface temperature, T is time and T_c is a reference time of $180/\pi$ days.

The depth of thaw is given by:

$$x_t = \sqrt{\frac{k_u t_{max} T_c}{L}} \left\{ 2 \sin\left(\frac{T}{2T_c}\right) - \frac{Ste}{3} \sin\left(\frac{T}{2T_c}\right) \sin\left(\frac{T}{T_c}\right) \right\} \qquad (2,24)$$

Although the prediction of the total depth of thaw will be similar using the time-dependent method and the

46

Neuman   step method, Nixon and McRoberts (1973) show that
the former predicts an initially slow penetration of the
thaw front, which subsequently becomes faster, whereas
the latter predicts a constant intermediate thaw rate.
The time-dependent model is likely to be the more
realistic.

With the introduction of computer modelling of ground
thermal regimes complex factors such as varying surface
conditions, spatially dependent thermal properties, and
temperature dependent thermal properties may be included,
using numerical methods to obtain a solution. These
numerical methods adopt finite difference or finite
element approaches to the problem of modelling temperature
and phase change in soil (Harlan and Nixon 1978). A
detailed discussion of such techniques is beyond the
scope of this book and the reader is referred to
Lachenbrook (1970), Ho et al. (1970), Hwang et al.(1972),
Outcalt (1972), Jahns et al.(1973), Goodrich (1974) and
Outcalt et al. (1975). Some results of a one-dimensional
finite difference computer program to illustrate the
effect of surface conditions on the ground thermal regime
are given by Gold et al.(1972) and Smith and Tvede (1977)
illustrate the use of Outcalt's model to predict soil
surface temperatures from inputs of weather data. The
surface temperatures are then used to predict the ground
thermal conditions using a one-dimensional finite
difference solution of the heat conduction equation. The
model accommodates complex stratigraphy and freeze-thaw
energetics.

For practical geomorphological and geotechnical
studies however, the simple Neuman / Stephan formula is
usually sufficiently accurate for predicting rates and
depths of soil freezing and thawing. The measurement or
calculation of the soil thermal regime is often an
important pre-requisite for an adequate quantitative
understanding of the behaviour of slopes subject to
periglacial mass movements.

3.   FROST HEAVE AND SOIL CREEP

Introduction

Observations show that when many soils freeze distinct
lenses of clear ice develop within them, parallel to the
ground surface. The resulting upward displacement of the
ground is termed frost heave. The relationship between
frost heaving and ice lensing in soils was first
recognised by Taber (1916). In later experiments (Taber
1929, 1930) he measured frost heave for different soil
samples using an open system whereby water was free to
enter the specimen as freezing took place. He found that
the measured uplift of the surface was greater than that
which would have resulted from *in situ* freezing of the
pore water in a saturated sample, and therefore concluded

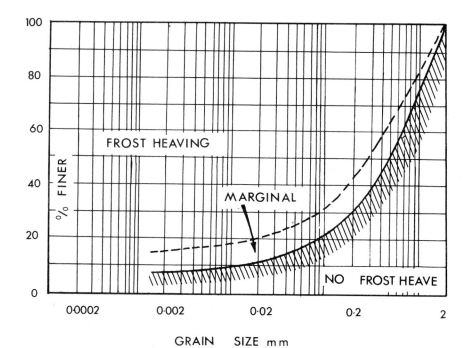

GRAIN SIZE mm

Figure 24 Beskow's textural limits of frost
susceptibility (Beskow 1935).

that water was drawn in from outside during the freezing
process. Taber showed that the major factors controlling
the amount of frost heaving were: size and shape of
soil particles, amount of water available for freezing,
size and percentage of voids, rate of cooling, and
surface load resisting heave.

Beskow (1936) also studied the freezing behaviour of
soils with varying particle size distributions. He
showed that frost heaving increases rapidly from nearly
zero for coarse sand to a maximum in the fine silt sizes,
from which it slowly declines to approach zero again in
heavy clay. For engineering purposes Beskow proposed a
division of soils into non-frost susceptible and frost
susceptible groups, and presented an empirically derived
grading such that soils with finer grain size were
defined as 'normally frost heaving', while soils with
coarser grain size were defined as 'not normally frost
heaving' (figure 24).

This may be simplified to a general statement that
coarse and medium sands are generally non-frost
susceptible, that is ice lenses do not normally develop
when they freeze, whereas fine sands, silts and all but
the heaviest clays are frost susceptible and are subject
to considerable ice lensing during freeze, providing a
water supply is present. The freezing of a fine grained

48

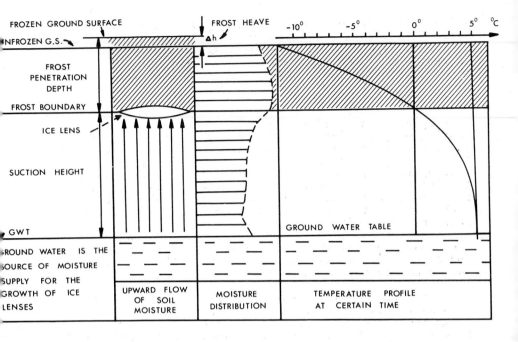

Figure 25   Freezing model for a fine-grained soil
             (Jumikis 1956).

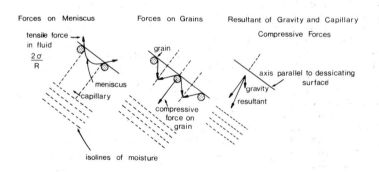

Figure 26   Capillary forces acting on an inclined plane
             parallel to a dessicating surface, and
             resultant considering gravity (Washburn 1967)

soil is illustrated diagrammatically in figure 25. Although all saturated soils will expand on freezing due to the 9% expansion of pore water accompanying phase change, where ice lenses do not develop the resulting heave is generally not great, and in the context of slope stability is of minor significance.

Davison (1889) stressed the importance of frost heaving as a cause of soil creep. He described the heaving of particles at right angles to the slope and their near vertical collapse following thaw, producing a net downslope displacement. Davison stated that on thawing the soil particles do not settle vertically under the influence of gravity, but somewhere between vertical and normal to the slope, because of the 'adherence of each particle to its neighbours, by reason of the water between them' which 'tends to bring it back to its old position' (p.259). His experimental results suggest that the departure from the vertical during settling approximately bisects the angle between the vertical and the normal to the surface of the soil.

Washburn (1967) discussed the nature of the forces tending to cause resettling of the soil particles following heave. He showed that true cohension occurs only in clays, but apparent cohesion, due to capillary water in silts produces a pressure normal to the slope and so causes departure from the vertical (figure 26). Capillary pressures due to surface tension can be many times as strong as the effect of gravity.

From Davidson's theory of the frost creep mechanism, attempts have been made to predict the profile of movement with depth resulting from frost creep. Williams (1957) shows that if all factors governing the formation of ice lenses are the same at all depths, for one freezing cycle the velocity profile in the soil due to creep will be a straight line (figure 27a). However, he points out that grain size, freezing rate, moisture supply etc. are not constant with depth and therefore the rate of heave may vary per unit of depth, giving a curved velocity profile of the form shown in figure 27b. Benedict (1970) suggests that the velocity profile due to creep is likely to be of the form shown in figure 27c, because the number of freezing and thawing cycles penetrating the soil increases towards the surface. Kirkby (1967) initially assumes that the rate of soil movement is proportional to the frequency of freezing and thawing times the expansion coefficient of the soil, giving a velocity profile similar to that of Benedict (1970). However, Kirkby points out that at the surface there is no overburden, so that the gravitational force there is zero. Therefore at the surface no downslope movement relative to the layer immediately below can take place, despite the fact that heave is maximum. At great depth the soil is not affected by expansion or contraction, so there can be no creep movements however

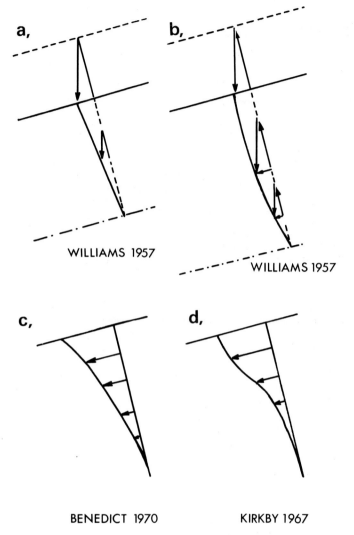

a,

WILLIAMS 1957

b,

WILLIAMS 1957

c,

BENEDICT 1970

d,

KIRKBY 1967

Figure 27   Theoretical velocity profiles due to frost
creep.

great the overburden pressure.   In between, at some
finite depth, will be a zone of maximum net creep
movement so that the velocity profile is qualitatively
of the form shown in figure 27d.

The amount of frost heaving of the soil on a slope
therefore directly influences the rate of creep
displacement.   In addition, the growth of ice lenses
during soil freezing draws water from below to the
freezing plane (figure 25).   These ice lenses disrupt the
soil structure, and in particular, increase the void
ratio.   On thawing the soil has excess water and an

extremely low bulk density, leading to saturation, consolidation, and consequent loss of strength (see chapter 4). The amount of frost heaving is therefore of considerable importance with respect to periglacial mass movements. Indeed it could be argued that the ice content of thawing soils is the most critical factor governing slope stability in periglacial regions.

Phenomena observed during soil freezing

During freezing of frost susceptible soils ice lenses generally grow parallel to the ground surface and perpendicular to the direction of heat flow. Such ice lenses correspond to Higashi's ice filament layer type of soil freezing (Higashi 1958). Although the ice lenses look like pure ice plates in the soil Higashi showed that they are in fact composed of bundles of numerous ice filaments of minute crystals. They contain air bubbles elongated in the vertical direction giving the appearance of columnar structure to the ice. A second form of segregation ice identified by Higashi is the sirloin type where thin ice layers are dispersed throughout the frozen soil giving it an appearance similar to a piece of sirloin beef. In non-frost susceptible soils and where frost susceptible soils are frozen very rapidly, ice segregation is restricted and the frozen soil has the appearance of concrete, leading Higashi to call this the concrete type of freezing.

The amount of frost heaving resulting from the freezing of a given soil may be represented by the Heaving Ratio, where:

$$\text{Heaving Ratio} = \frac{\text{frost heaving at the surface}}{\text{thickness of the frozen layer}}$$

which can be written as:

$$H = \frac{dh/dt}{dx/dt} \qquad (3,1)$$

where H is the heaving ratio, dh/dt is the rate of frost heaving and dx/dt is the rate of penetration of the freezing plane. Since it is assumed that frost heaving is the result of the growth of ice lenses (rather than expansion on freezing of *in situ* pore water), it is possible to estimate the excess water content of the frozen soil layer, since, per unit volume of soil;

$$W_{ex} = \gamma_i H \qquad (3,2)$$

where γ_i is the density of ice, H is the heaving ratio, and W_{ex} the excess water content of the frozen soil.

The flow of water to the freezing zone from the unfrozen soil below in frost susceptible soils is the result of a suction gradient set up in the soil water as a

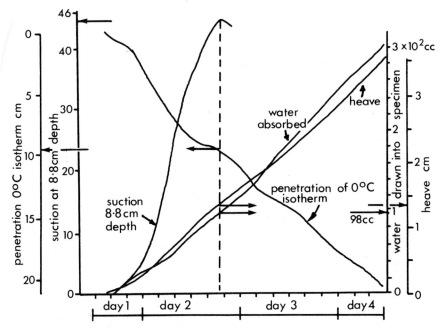

<u>Figure 28</u>   Laboratory soil freezing experiment (Jumikis 1956).

result of ice crystallisation (Jumikis, 1956, Penner 1959, Williams, 1977).

Jumikis (1956) monitored the freezing of a soil sample held in an open system such that water could enter the sample at the base during the freezing process. Data presented illustrate the temperature change, and negative pore water pressure (suction) at a depth of 9 cm (3.5 in) in the sample, (figure 28), and show that the suction reached a maximum of 33.3 cm mercury (44.6 kgm^{-2}) at the point where soil freezing at this depth began. This suction value occurred at a temperature of -0.266°C (31.5°F), suggesting that soil freezing commenced at a temperature below 0°C in this soil. In freezing from the surface to a depth of 9 cm the soil specimen drew in 98 cm^3 water from the external source, and the surface was heaved by approximately 13 mm (0.53 in).

Williams (1966, 1968, 1972) describes the laboratory measurement of pore water pressure at the freezing line, and figure 29 shows pore pressures recorded at the base of small cylindrical soil specimens subject to freezing from above. The specimens were placed on a saturated porous filter which was connected to a pore water pressure measuring device. The value u_i in figures 29a and 29b is the pore water pressure when the freezing plane reached the base of the soil sample, immediately above the filter, and is therefore the pore pressure at the frost line.

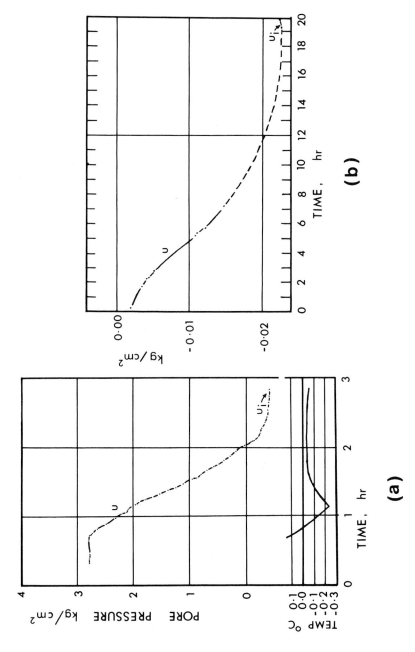

Figure 29   Pore pressures measured during penetration of frost line in (a) Leda Clay, and (b) 73%-75% silt fraction (Williams 1966).

Water will only flow to the frost line if the pore pressure there is lower than the pore pressure below, in other words, if a pressure gradient towards the frost line is generated in the unfrozen soil below.

The freezing of soil water

It has long been established that water in soil pores may not freeze at $0^{\circ}C$, but at some lower temperature. Early investigators, noteably Bouyoucos (e.g. Bouyoucos 1913 and 1916) suggested that solutes dissolved in soil water are responsible for the freezing point depression of soil water.

Later investigators, including Taber (1929, 1930) and Beskow (1935) maintained that of more general importance is the interaction between the soil solids and soil water. Molecular attraction at the interface between the soil particles and the soil water causes an 'adsorbtion pressure' in the water film in contact with the soil particles, producing a so-called 'adsorbed layer' in which the freezing point is depressed. Freezing point depression is proportional to the adsorbtion pressure, which decreases logarithmically with distance from the surface of the solid (Winterkorn and Baver 1934). Therefore in the outer layers of the adsorbed film where water is only slightly affected by adsorbtion pressures freezing occurs near $0^{\circ}C$, but as the particle surface is approached progressively lower freezing temperatures are required to overcome the increasing adsorbtion pressure within the film. A soil which undergoes a steady cooling below $0^{\circ}C$ will therefore contain ice and water in its pores, the proportion of water decreasing with falling temperature, as more of the adsorbed water freezes and the adsorbed layer thins. In fine grained soils the pores may be so small that all the pore water is affected by adsorbtion pressures. Soil freezing in this case will not begin until the temperature has fallen below $0^{\circ}C$.

Another approach to the problem of the freezing point depression in soil pore water is based on principles of thermodynamics, and is susceptible to mathematical treatment. The theory is outlined by Williams (1968), and begins with the consideration of pressure differences between a small spherical ice crystal and water in which it is submerged, given by:

$$p_i - p_w = \frac{2\sigma_{iw}}{r_{iw}} \qquad (3,3)$$

where p_i = pressure of ice
p_w = pressure of water
σ_{iw} = surface tension ice-water
r_{iw} = radius of interface.
It is apparent that the pressure difference between ice and water increases as the radius of curvature of the interface decreases. This pressure difference influences

55

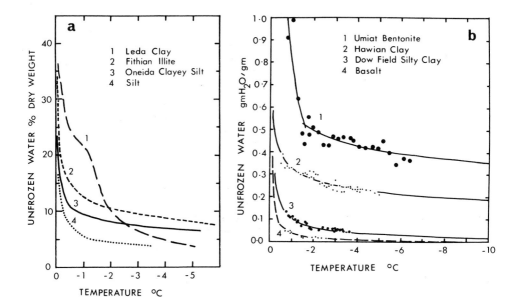

Figure 30   Unfrozen soil water content as a function of
            temperature , (a) from Burt and Williams 1976,
            and (b) from Anderson and Morgenstern 1973.

the freezing point, which falls as the pressure difference
increases.   These pressure relationships may be related
to the freezing point of soil pore water by means of the
equations:

$$\log \frac{T}{T_O} = \frac{V_1 2\sigma_{iw}}{r_{iw}L_w} \qquad (3,4)$$

$$\text{or} \qquad T - T_O = \frac{V_1 2\sigma_{iw}T_O}{r_{iw}L_w} \qquad (3,5)$$

where T = freezing point ^{0}K
      T_O= normal freezing point (i.e. when the pressure
         is uniform on the two phases).
      V_1= specific volume of water
      L_w= latent heat of fusion of water.
When a soil is cooled below 0^0C ice formation normally
begins at the surface at a temperature close to 0^0C and
progresses downwards initially through the larger pores,
where, because the radius of curvature of the advancing

56

ice is larger, pressure differences between ice and water are less and freezing point depression is least. As the temperature falls ice is able to penetrate into smaller openings and spread through the soil. Because in any given soil there is usually a range of pore sizes, and consequently a range in the radii of curvature of ice-water interfaces during soil freezing, unfrozen water can exist in the finer pores at temperatures below $0^{\circ}C$ when the larger pores are ice filled, and freezing and thawing of soil occurs over a range of temperatures. A similar description of the soil freezing processes is given by Penner (1959).

The presence of unfrozen water in frozen soils is widely reported and diagrams illustrating the relationship between temperature and unfrozen water content are given in figure 30. Methods of measuring the unfrozen moisture contents of frozen soils are described by Anderson and Tice (1973) and Anderson and Morgenstern (1973).

The nature of the unfrozen water present in frozen soils has been investigated by Anderson (1967) by means of X-ray diffraction studies of frozen montmorillonite-water mixtures. Observations show that an essentially liquid-like, unfrozen interfacial layer of water separates ice from the silicate surface of the soil particles. The thickness of this interface is shown to range from more than 15Å at $0^{\circ}C$ to about 6Å at $-5^{\circ}C$ in the montmorillonite-water mixture.

A simple demonstration of the presence of a thin film of water between soil ice and soil particles is provided by Corte (1962) who froze water from below, and sprinkled small soil grains onto the upward advancing ice front. The finer grains were observed to move up with the ice front, indicating that they were supported by a liquid-like film (not visible or measurable in the experiment) for otherwise the water would have frozen around them and they would have been engulfed by the ice.

Ice segregation and frost heaving

According to the adsorbtion theory ice segregation takes place in fine-grained soils because the water occupying their fine pores is subject to adsorbtion pressures, and therefore does not freeze at $0^{\circ}C$, but at some temperature below this. However, such soils also contain occasional larger pores where free water will freeze at $0^{\circ}C$. Thus at the $0^{\circ}C$ isotherm progresses through the soil, freezing of soil water is initially concentrated in the large pores where ice lenses grow by drawing up water from the unfrozen soil below. These lenses are separated from the adjacent mineral grains by the adsorbed water film. The process whereby water is drawn to the developing ice lens is explained by the need to maintain the thickness of the adsorbed water layer between the ice lens and the surrounding soil particles.

As molecules from the adsorbed film are incorporated into the growing ice lens at a particular temperature, they are replaced by molecules of water from the unfrozen soil below, in order to maintain the equilibrium thickness of the adsorbed layer at that temperature.

The thermodynamic theory also predicts that ice lenses develop initially in the larger pores where the ice-water interfaces have a relatively large radius of curvature. The pressure difference between the ice and water at the interface (equation 3,3) generally produces a fall in pore pressure at the interface so that water is drawn to the interface from below and supplies the growing ice lens. As long as the pore water pressure in the unfrozen soil below the ice lens remains higher than the pore pressure at the ice water interface the ice lens will continue to grow. However, the migration of water to the lens leads to a fall in pore pressure until the pressure in the finer pores is equal to that at the ice-water interface. This then allows the ice to advance through the finer pores and terminates the growth of the lens. This situation is described by the equation:

$$P_i - P_w = \frac{2\sigma_{iw}}{r_c} \qquad (3,6)$$

where r_c = radius of curvature of the ice water interface in passing through the finer pores (Williams 1968).

When ice lens growth has been terminated the pore pressures usually begin to rise again so that the condition for development of an ice lens will again rise. 'It is for this reason that frost heaved soils normally consist of alternate layers of frozen ground and ice lenses' (Williams 1968, p.95).

Penner (1959) similarly refers to the radius of curvature of the ice water interface controlling freezing point depression, and therefore leading to ice lens growth in the larger pores. However, he explains water movement from below to the growing ice lens as resulting from the need to maintain an equilibrium thickness of the adsorbed water layer between the ice lens and the soil particles. Kaplar (1970) stresses the role of a film of water separating growing ice lenses and soil particles in the frost heave mechanism, and suggests that the maximum heaving action will occur when the area of active film layers is maximum in a freezing front system. Similarly Takagi (1978, 1979) emphasises the importance of an adsorbed film separating ice and soil mineral grains, and suggests that the suction which draws water to the freezing front is generated by the need to maintain an equilibrium thickness of this film. Takagi in fact questions the validity of the capillary theory based on pressure differences across an ice/water maniscus (equation 3,6) since it is based on static conditions, rather than the dynamic condition of water flow and ice

formation which obtain in freezing soils. He suggests
that a more satisfactory approach is to consider ice
segregation as resulting from simultaneous flows of heat
and water.

Anderson and Morgenstern (1973) describe the ice
lensing processes in general terms, explaining that ice
lens growth occurs at locations where the temperature
is favourable and the rate of appearance and dissipation
of latent heat does not exceed the upward flux of water.
Growth continues until the balances between the four
principal governing factors, the soil texture, the rate
of heat removal, the upward flux of water, and the
confining pressure, are disturbed. The energy released
on freezing is partly converted into the work of lifting
the overburden and the remaining energy is dissipated as
heat at the freezing interface. At some point the upward
heat flux required by the thermal boundary conditions may
exceed the capacity of the soil water to supply latent
heat. When this happens, the freezing interface descends
to a new plane where ice formation and growth begins
anew.

Factors influencing frost heave

1) Texture

The importance of soil texture in the susceptibility
of soils to ice lensing has been mentioned earlier, and
Beskow's textural criteria illustrated (figure 24).
Jackson and Chalmers (1958) suggest a rough criterion
that soils containing more than 3% particles finer than
0.02 mm exhibit frost heaving, although they stress that
factors such as particle shape and chemical composition
may influence the susceptibility of a particular soil to
frost heaving.

Kaplar (1970) measured the rate of heave under
laboratory conditions for two gravelly sands, both
initially containing about 5% finer than 0.02 mm material,
with maximum grain size about 5 cm. Several specimens
were prepared and progressively more of the coarser
fraction removed, so that the percent finer than 0.02 mm
was increased. No other property of the specimens was
altered except the unit dry weights which differed, due to
the change in texture. Figure 31 shows heave rates
plotted against percent finer than 2 mm. Clearly texture
is of major significance in frost susceptibility,
particularly the proportion of fines present in the soil.

Closely related to grain size is pore size, and
according to the thermodynamic theory of frost heaving
pore size is the main factor controlling the
susceptibility of a soil to frost heaving, given adequate
moisture supplies. Csathy and Townsend (1962) describe
the measurement of pore size using the capillary rise of
water through samples enclosed in vertical tubes. They

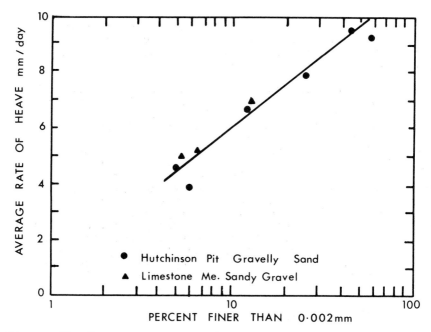

Figure 31 Heave rate plotted against percent finer than
           0.02 mm size of 'scalped' gradations (Kaplar
           1970).

propose a pore size distribution criterion for assessing
frost susceptibility and show that it compares favourably
with earlier textural criteria including those of Beskow
1935. Williams (1966, 1967, 1968, 1972) also stresses
the importance of pore size in the assessment of frost
susceptibility, and describes a method of measuring pore
size indirectly by means of the intrusion of air into
saturated soil specimens. The pressure required for air
to spread through the soil, displacing the pore water,
depends on surface tension effects, according to the
equation:

$$P_a - P_{wa} = \frac{2\sigma_{aw}}{r} \qquad (3,7)$$

where P_a = air pressure
      P_{wa}= water pressure (atmospheric)
      σ_{aw}= surface tension air-water
and r = size of the largest continuous openings through
the soil system. The air intrusion test therefore
provides a method of estimating r, and has been shown to
give an accurate assessment of soil frost susceptibility.
The method has the advantage that it can be used in the
field.

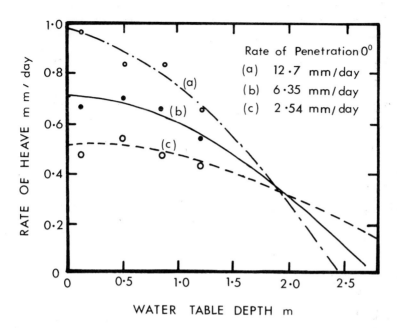

Figure 32   Influence of water table depth on rate of frost
heaving in New Hampshire silt (McGaw 1972).

2) Permeability and moisture supply

The growth of ice lenses in a soil during freezing
depends upon the relationship between the rate of heat
extraction from, and the rate of flow of water to, the
freezing plane. If the rate of water supply is
insufficient, the freezing plane will penetrate downwards
and ice lensing will be restricted. This explains the
widely observed reduction in frost heaving clays where
permeability below the freezing plane is too low to
maintain sufficient water flux to the freezing plane. Ice
lenses therefore tend to be thin.

Moisture supply during soil freezing is therefore a
major factor influencing the amount of ice segregation
which actually occurs in the field. McGaw (1972) in a
series of laboratory experiments investigated the
influence of depth to the water table on the rate of frost
heave for given rates of frost penetration. His results
clearly illustrate that frost heaving is reduced as the
depth to the water table is increased (figure 32).

Penner (1959) stresses that ice lensing may not
necessarily be associated with saturated soils, and
therefore saturated hydraulic conductivity. If the soil
below the ice front is not saturated, its hydraulic
conductivity is a function of moisture content, or more
precisely, pore water tension, or pore water suction
(figure 33). As would be expected, the unsaturated

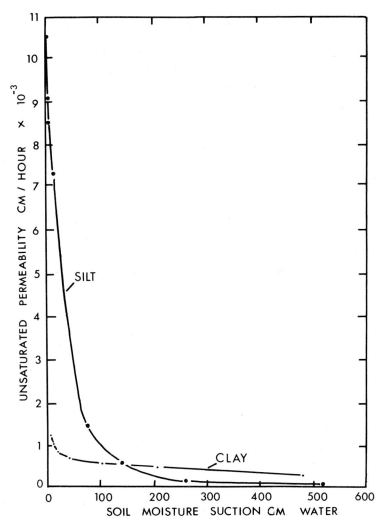

Figure 33 Unsaturated permeability and suction-moisture content relationships (Penner 1959).

hydraulic conductivity, and therefore the supply of water to a growing ice lens, falls as the moisture content of the soil falls.

Jumikis (1973) investigated experimentally transfer of water from the groundwater to a freezing plane by unsaturated flow of a non-Darcian nature. He suggested that such flow is via the water films wetting soil grains for porosity values up to approximately 60%. Above this the much less efficient vapour diffusion mechanism occurs. The rate of water film flow was found to increase with increasing porosity, up to a value of about 42%, above which transfer rates fell.

The mechanism of frost heaving in unsaturated soils is discussed in detail by Miller (1972), who refers to the process as secondary heaving. Because the transfer of water to the freezing plane via water films is much slower than Darcian flow in saturated soils, secondary heaving requires a much longer time-scale and in practice is much less important than primary frost heaving of saturated soils.

Young and Osler (1971) point out that at low moisture contents soil freezing is associated with ice expanding into air filled pores, and no heave therefore takes place. At higher moisture contents, water and subsequently ice, completely fills some pores, and heaving becomes possible.

Secondary heaving may be a significant mechanism for increasing the ice content of already frozen soil above the freezing line. The presence of unfrozen water as films separating the ice and soil particles in frozen soils enables water film flow to occur and supply the continued slow growth of ice lenses behind the freezing front. The hydraulic conductivity of frozen soils has been measured by Burt and Williams (1976) at temperatures a few tenths of a degree below zero, and are shown to decrease rapidly to between 10^{-10} and 10^{-8} cm per second as the temperature falls to around -0.2 to -0.3°C, and to decrease more slowly below this temperature. Anderson and Morgenstern (1973) suggest that at temperatures below about -5°C the thinness of the unfrozen interface severely restricts the quantities of water that can be transported, although in theory ice lens growth can continue until about -50°C. Secondary heaving may be significant in increasing the ice content of permafrost, where long time scales are available for moisture transfer to take place.

3) Rate of heat extraction

Since the growth of ice lenses depends on the relationship between the removal of latent heat and the flow of water to the freezing front, the rate of heat extraction from the soil is clearly an important factor influencing frost heave. It is important to distinguish between the rate of heave and the heave per unit depth of frozen soil (heaving ratio), since the former is directly related to the rate of frost penetration (Penner 1959, 1960, 1972, McGaw 1972) and the latter is inversely related to the rate of frost penetration (Taber 1929, 1930, Williams 1968, McGaw 1972).

As Penner (1972) has shown, more rapid heat extraction from the soil leads to more rapid freezing, more rapid flux of water to the freezing front, and consequently more rapid frost heaving. However, Kaplar (1968) shows that at very rapid rates of freezing the heave rate begins to decrease with increases in the freezing rate and eventually it intersects the heave rate due simply to the

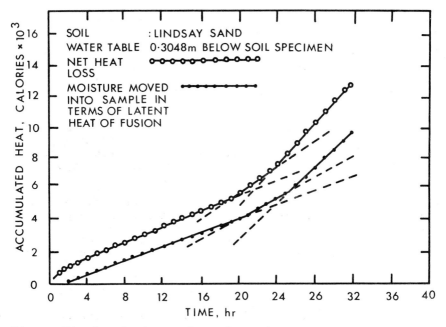

Figure 34   Cumulative value of net heat flow and moisture
flow versus time during laboratory freezing
of soil (Penner 1972).

expansion of water freezing *in situ* in the soil voids.

Under these conditions (unlikely to be met in the
field) freezing is too rapid to allow water to be drawn
to the freezing front from below. Horiguchi (1979) shows
similar results in laboratory tests of various powders.

The reduction in the heaving ratio (and ice content
per unit volume of frozen soil) with increased rate of
frost penetration is explained by the reduction in time
available at any given depth for ice lens growth. This is
well illustrated by a consideration of the ice
segregation efficiency ratio introduced by Arakawa
(Arakawa 1966). This is the ratio of the rate of latent
heat produced at the freezing front (which is a measure
of the rate of water flow to the front from below) to the
rate of heat flow from the soil, expressed by the
following equation:

$$E = \frac{di}{dt} L_w / [k_f (\frac{dT_f}{dx}) - k_u (\frac{dT_u}{dx})] \quad (3,8)$$

where $\frac{di}{dt}$ = rate of ice segregation;

L_w = latent heat of fusion;
k_f = thermal conductivity of frozen soil;
k_u = thermal conductivity of unfrozen soil;

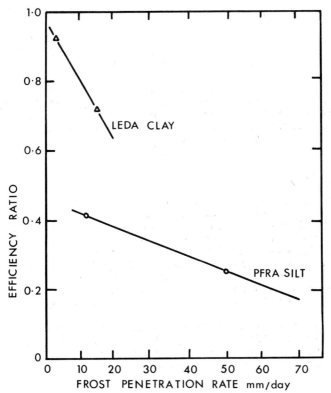

<u>Figure 35</u>   The ice segregation efficiency ratio as a function of frost penetration rate (Penner 1972).

$\dfrac{dT_f}{dx}$ = thermal gradient in frozen layer;

$\dfrac{dT_u}{dx}$ = thermal gradient in unfrozen layer.

In laboratory studies Penner (1972) has measured rates of heat extraction and rates of moisture drawn into freezing samples under different rates of frost penetration (figure 34). The calculated ice segregation efficiency ratios against frost penetration rate are shown in figure 35. For Leda clay increasing penetration rates lead to a rapid fall in efficiency ratio, due to the low permeability of the clay and consequent slow movement of water to the freezing front from below. The fall in efficiency ratio with increasing frost penetration rate for the silt is less rapid, reflecting its higher permeability.

When the ice segregation efficiency ratio equals 1 the rate of latent heat production at the freezing line equals the rate of heat extraction from the soil (ignoring the small amount of heat needed to cool water as it moves up through the unfrozen soil the freezing line) so that equation 3,8 becomes:

$$k_f \left(\frac{dT_f}{dx}\right) - k_u \left(\frac{dT_u}{dx}\right) = \frac{di}{dt}L_w \qquad (3,9)$$

and frost heaving takes place with a stationary frost line. This condition is referred to as 'perfect segregation' by Arakawa. Imperfect segregation occurs when $0<E<1$ and the freezing front advances through the soil and for a soil which does not heave $E = 0$, since $\frac{di}{dt} = 0$.

4) Surface load

When a surface load is applied to a freezing soil frost heaving is resisted, and the amount of heave reduced. In order to overcome the surface load heaving pressures must develop in the soil. When heaving is prevented completely considerable heaving pressures may be generated. Penner (1970) in a field study of heaving pressures developed in leda clay recorded a maximum force developed under a 30.5 cm diameter steel plate in excess of 13608 kg, giving a heaving pressure of 46558 kg/m².

From a consideration of thermodynamic principles, Everett and Haynes (1965) develop the following equation for heaving pressure of a continuous ice lens. In an ideal close-pack array of spherical beads:

$$\Delta P = \frac{2\sigma_{iw}}{r_i} + \frac{2}{r}(\sigma_{is} - \sigma_{ws}) \qquad (3,10)$$

where ΔP = heaving pressure
σ_{iw}= ice-water interfacial energy
σ_{is}= ice-particle interfacial energy
σ_{ws}= water-particle interfacial energy
r_i = radius of pore
r = radius of particle

The first component on the right hand side of equation 3,10 is equal to the pressure drop across the curved ice-water interface; the second is referred to as the 'flotation effect', as demonstrated by Corte (1962). It is apparent that theoretical heaving pressures increase with decreasing particle and pore sizes.

Sutherland and Gaskin (1973) show by laboratory tests on soils with various textures held in closed systems, that an increased percentage finer than 0.02 mm leads to a greater drop in pore pressure during freezing (increased pore water suction) and the development of increased heaving pressures. Penner (1968) in open system laboratory tests on various fractions of a silt soil also demonstrates that heaving pressures increase as particle

66

size decreases, and he stresses the importance of the finest fractions in a given soil in generating frost heave and heaving pressures. Young and Osler (1971) in a series of open system experiments show that the heaving pressure developed at the soil surface increases as the force restraining surface heave is increased, but the amount of heave at the surface decreases as the restraining force is increased.

Under natural conditions frost heaving is resisted by the weight of frozen soil above the freezing front; that is the overburden pressure. In view of the high heaving pressures which frost susceptible soils are capable of generating, frost heave in the active layer of seasonal freezing and thawing is likely to be little affected by overburden pressure.

Williams (1966, 1968) demonstrates the importance of overburden pressure in the development of ice lenses in marginally frost susceptible soils. Frost heave can only occur if the pore pressures developed at the freezing front are lower than those in the soil below, so that a hydraulic gradient exists and water can flow to the freezing soil. The pore pressure at the penetrating frost line can be predicted from the measured air intrusion value of the soil since:

$$(P_i - P_w) = A.\sigma_{iw}/\sigma_{aw} \qquad (3,11)$$

where P_i = pressure in ice
P_w = pore pressure at the ice front
A = air intrusion value (equals P_a-P_{wa} in equation 3,7)
σ_{iw}/σ_{aw} = ratio of surface tension ice-water and surface tension air-water.
Therefore:

$$P_w = P_i - A.\sigma_{iw}/\sigma_{aw} \qquad (3,12)$$

If frost heaving occurs, the overburden pressure must be supported by the ice, so that $P_i = \gamma x$, where γ is the bulk density of the frozen soil, and x is the depth of freezing. The pore pressure at the freezing line, P_{wx} is therefore defined as a function of depth, since:

$$P_{wx} = (\gamma x - A.\sigma_{iw}/\sigma_{aw}) \qquad (3,13)$$

where P_{wx} is the pore pressure at the ice front at a depth x from the surface. The pore pressure in the soil below the freezing plane depends on depth below the surface and location of the water table. Assuming hydrostatic conditions prevail, if u_x is the pore water pressure before freezing at a depth x,

$$u_x = \rho(x - z) \qquad (3,14)$$

where ρ = density of water

z = depth to water table.

Water migration to the frost line to feed the growth of ice lenses occurs only when $u_x > P_{wx}$. From equation 3,13, P_{wx} is proportional to $\gamma.x$, and from equation (3,14), u_x is proportional to $\rho.x$, and since $\gamma > \rho$, P_{wx} must increase more rapidly with depth than u_x. There must therefore be some depth below the surface where $P_{wx} = u_x$ at which water migration to the freezing plane and ice segregation ceases. In soils which are only slightly frost susceptible this may be only a few decimetres below the surface (Williams 1966). Williams (1968b) used the theory developed above to explain the distribution of ice lenses in permafrost in the Mackenzie Delta. Although variation in soil texture and moisture supply with depth introduced some irregularity in the profiles examined, generally ice content in the frozen soils was greatest near the surface, and in two of the profiles little water migration and ice accumulation had occurred during freezing, except in the near surface layers, where substantial accumulation had taken place.

Field studies of frost heaving

The majority of frost heave studies in the field have been made in areas of patterned ground rather than on slopes.

The most common technique for field measurement of frost heave is by means of plates on the ground surface or buried at various depths below the surface, connected by moveable rods to a fixed frame, the legs of which are anchored in permafrost, or protected from frost heaving by sheathing and burial in non frost-susceptible gravel. Upward displacement of the plates by frost heave may be measured by the corresponding movement of the rods through the frame (Czeppe 1959, 1960, Heywood 1961, Matthews 1967, James 1971, Harris 1972 and Fahey 1974). Refinements include the continuous recording of rod movements by means of a pen recorder (James 1971, Matthews 1967 and Fahey 1974).

Washburn (1967) measured frost heave, resettlement, and lateral downslope soil movements by theodolite survey of target cones, and Benedict (1970) measured frost heave of the ground surface against a tensioned steel wire strung along a line of stakes between benchmarks consisting of heavy steel pipes. These were inserted below the depth of frost heaving, cemented in place, and protected from heaving by greased metal casing.

Everett (1966) used linear motion transducers to record frost heaving in Alaska. Movement of the soil against a small plate connected to the transducer by a short aluminium shaft was recorded as a change in electrical resistance on a continuously recording Wheatstone bridge. The transducer was attached to an

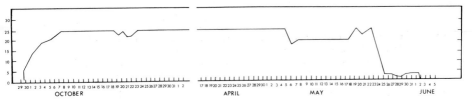

Figure 36   Frost heaving in a stony mud circle, Cape
Thompson, Alaska 1961-62 (Everett 1966).

anchor set in permafrost to ensure its immobility.
Similar transducers were attached to a frame apparatus by
Giles (1973) so that heaving of the surface against
pressure plates was transmitted via rods and recorded by
a battery driven electrical recorder which scanned each
rod once every hour.

For detailed descriptions of apparatus used in frost
heave studies the reader is referred to the survey by
James (1971).

The major factors influencing frost heaving in the
field include soil texture, water supply, soil thermal
regime and depth of freezing and thawing.

Benedict (1976) compared annual frost heave
measurements by various workers in regions of shallow
seasonal freezing, deep seasonal freezing, and shallow
permafrost or bedrock. He showed that heaving is
greatest where freezing is deep and where it continues
throughout the winter. Shallow permafrost or bedrock
reduces the thickness of the layer experiencing ice
segregation and therefore reduces surface heave.

The influence of textural contrasts between the
coarser perimeter and the fine grained centre of a sorted
circle in Signy Island, Antarctica was illustrated by
Chambers (1967). During the winter of 1964 he recorded
total heave values of between 40 and 47 mm in the central
area but 13 mm and less adjacent to the perimeter. Frost
heaving was also only recorded during freezing of the
upper 40 cm of soil. When the freezing plane had
penetrated below this depth no further heave was recorded.
The active layer was between 120 and 150 cm thick at this
site. In Spitsbergen, Czeppe (1960) found a greater
contrast between heaving of the fine grained centre of a
sorted circle and the coarser perimeter, with maximum
heave recorded in the former of 155 mm, and in the latter,
15 mm (1957-58).

Everett (1966) in a study of various types of
patterned ground in Alaska observed much greater frost
heave in stony and mud circles than over turf banked
terraces, heave in the autumn of 1961 being 25 mm
(pressure plate 9.1 cm below surface) and 32 mm (pressure
plate 4.0 cm below surface) in two nonsorted circles,

69

Table 8   Frost heave, Okstindan Mountains, North Norway
1969-70 and 1970-71.

	Winter Heave	Summer Settlement	Winter Heave	Summer Settlement
	1969-70 cm	1970 cm	1970-71 cm	1971 cm
Site A	2.8	3.5	9.1	8.8
Site B	0.7	1.8	2.5	2.1

Average rate of penetration of the freezing plane Site A 1970-71
1.23 cm/day
Average rate of penetration of the freezing plane Site B 1970-71
2.33 cm/day

compared with 0.4 mm (pressure plate 10.2 cm below
surface) in a turf banked terrace. The amounts of heave
recorded were found to depend on texture, moisture
content and the rate of freezing. Everett reports that
heaving began when ground temperatures reached a critical
value usually below zero. The rate of heaving was then
greatest where moisture contents and freezing rates were
high, and least where moisture contents were low and
freezing rates slow. Figure 36 shows heave data for a
stony mud circle for the winter 1961-62. Both heaving
during freeze and settlement during thaw were clearly
rapid, and some heaving associated with short period
freeze-thaw was recorded in the spring of 1962, following
initial thawing of the ground at this site.

Harris (1972a, b) investigated the relationship
between freezing rates and the amount of heave on
adjacent turf-banked solifluction lobes in Okstindan,
Norway. The textural properties (silty fine sand) and
soil moisture contents at the two monitored sites were
essentially the same. However in Site A snow accumulated
during autumn and winter to a depth of around 1 m, while
in Site B deflation prevented snow accumulation, and
maximum depths of 2.5 cm only were recorded. The result
was rapid soil freezing in the snow free site, and much
slower freezing in the accumulation site. Table 8
presents data for the winters of 1969-70 and 1970-71.

It may be concluded that the slower rate of freezing
in Site A allowed greater migration of water to the
freezing plane generating a higher heaving ratio and ice
segregation ratio.

Fahey (1974) measured frost heaving over a frost boil
on the surface of a large turf banked solifluction lobe
on the Niwot Ridge, Colorado Front Range. A recording
frost heave frame apparatus was used. Figure 37 shows

70

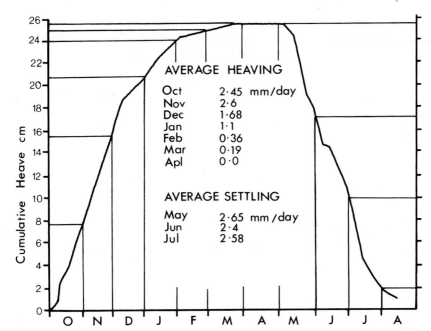

AVERAGE HEAVING

Oct	2·45 mm/day
Nov	2·6
Dec	1·68
Jan	1·1
Feb	0·36
Mar	0·19
Apl	0·0

AVERAGE SETTLING

May	2·65 mm/day
Jun	2·4
Jul	2·58

Figure 37   Frost heaving and resettlement at the surface
             of a frost boil, Saddle Site, Niwot Ridge,
             Colorado Front Range 1969-70 (Fahey 1974).

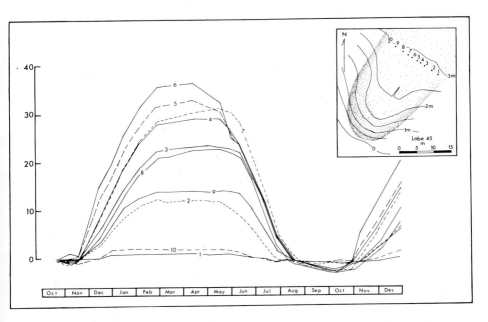

Figure 38   Summary of frost heaving 1965-66, Lobe 45,
             Niwot Ridge, Colorado Front Range (Benedict
             1970)

heave and subsidence rates during freeze and thaw at this
site. Comparison of frost penetration rates and rates of
frost heave at the surface indicates that higher rates of
frost penetration gave higher rates of frost heave. The
correlation between frost penetration rate and frost
heave rate is inconclusive, giving a coefficient of 0.557,
but it appears to have been lowered by the general
reduction in the rate of heave at greater depths during
the later part of the winter, despite consistently
moderate rates of frost penetration.

Perhaps the most useful study of frost heave in the
context of periglacial slope stability is that of
Benedict (1970) who monitored the processes acting in
turf banked solifluction lobes on Niwot Ridge, Colorado
Front Range. At one site heaving was recorded adjacent
to a series of pegs inserted across a small lobe (figure
38). Heaving was greatest in the central area and least
towards the sides of the lobe. This corresponded to the
moisture distribution, the lobe being saturated along its
axis but drier towards the sides.

The conclusions of McGaw (1972) concerning the
influence of depth to the water table on ice segregation
are supported by field data from Southern Ontario,
Canada (Fahey 1979). Frost heave in a frost susceptible
till was monitored near Guelph over the winters of 1975-
76 and 1976-77, and marked contrasts were observed.
Despite deeper frost penetration during 1976-77 frost
heaving was less than half that in 1975-76 (figures 39
and 40). During the winter of 1975-76 the water table
was less than 1 m below the surface when freezing began
in December and remained within 150 cm of the surface
throughout the winter. During 1976-77 however the water
table was more than 2 m below the surface, the restricted
ice segregation resulting from this reduced water supply
being well illustrated. As in studies elsewhere the
heave records show rapid heaving in early winter followed
by fairly stable conditions prior to rapid settlement in
spring. In February 1976 settlement of the ground
surface was due to thawing at the base of the frozen soil
rather than at the surface. Surface thawing in March
1976 was accompanied by very rapid resettlement.

Penner (1970) examined the heaving behaviour of Leda
clay in the field when subject to freezing to a depth of
approximately 84 cm (figure 41). The amount of heave at
various depths was recorded weekly (figure 42), and showed
that the commencement of heaving corresponded closely with
the time the 0^0C isotherm intercepted the gauge. The
vertical distance between gauges remained almost constant
after the frost line passed their positions. Slight
increases in these distances indicate some continued ice
segregation at temperatures below 0^0C, but it can be
concluded that the main heaving activity was at the
freezing plane. Heaving pressure developed against an
anchored plate was also measured, and reached a maximum

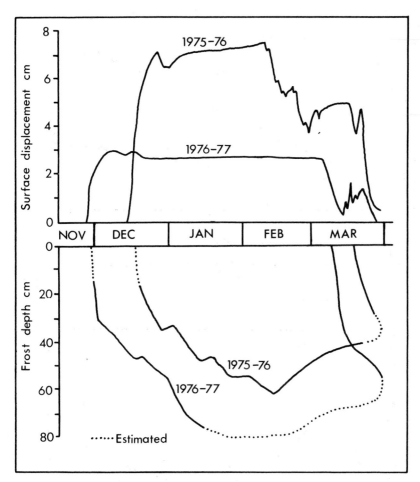

Figure 39   Relationship between surface heave and frost
             depth , 1975-76 and 1976-77, Guelph, Ontario
             (Fahey 1979).

in mid February 1969, before the maximum surface heave
displacement occurred. Pressures fell rapidly as soon as
thawing at the base of the frozen soil began in February.
Surface thawing began in early March and marked the
beginning of settlement at the surface. Clearly no ice
lenses had developed during the freezing of the deeper
layers of soil, since basal thawing was not associated
with consolidation during the second half of February
and the beginning of March. The maximum rate of
settlement during the spring thaw was approximately 1.8
mm/day at the surface in April.

    Andrew's analysis of frost heave data collected by
B.H.J. Haywood from Schefferville, Labradore-Ungava
(Andrews 1963) reveals rapid resettlement of the ground
during the thaw period, the ground surface returning to

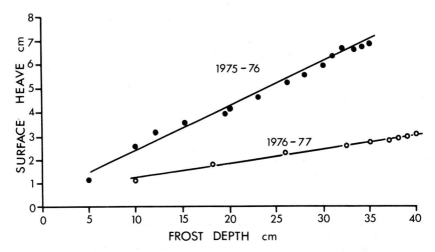

Figure 40   Relationship between frost heave and frost
depth during snow-free period at the beginning
of 1975-76 and 1976-77 seasons, Guelph Ontario
(Fahey 1979).

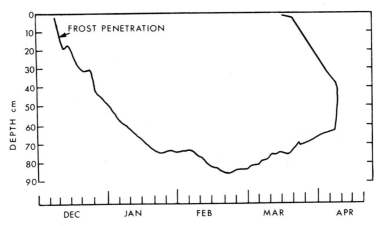

Figure 41   Frost penetration, Leda clay, 1968-69, Ottawa
(Penner 1970).

its pre-winter level in a few days. Andrews stresses the
importance of snow cover in prevented thawing of the
soil. Even under a snow cover of 6 mm the ground was
solidly frozen. Four main periods were identified during
the winter of 1959-60; late August to mid October, when
soil heaving was due to needle ice; mid October to
December, the main period of heaving as the soil froze to
bedrock (37 cm to 60 cm deep); January to May, no
movement; mid May, sudden and rapid collapse of the soil
as ice lenses thawed. Andrews also points out the lack
of correspondance between the frequency of freeze-thaw
cycles in the screen and in the ground. At Schefferville

74

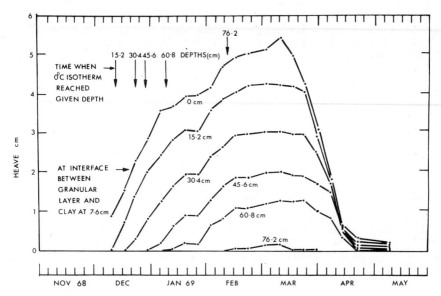

Figure 42   Frost heaving of Leda clay 1968-69, Ottawa
            (Penner 1970).

30 cycles were recorded in the screen between September
1959 and June 1960 but only two soil fluctuations
occurred in the autumn, and none in the spring.

   Diurnal freeze-thaw cycles and frost heave cycles in
the Indian Peaks region of the Colorado Front Range were
monitored by Fahey (1973) along an altitudinal transect.
He shows that the number of diurnal freeze-thaw cycles
decreased from 238 at 2,600 m to 89 at 3,750 m over a
twenty-two month period. At sites with persistent snow
cover however, diurnal freezing and thawing of the ground
was severely restricted, and at only one snow free site
(Ponderosa, 2200 m) was the number of frost heave cycles
significantly correlated to the number of diurnal freeze
thaw cycles. In all sites many of the recorded frost
heave events in fact resulted from the action of needle
ice at the soil surface, particularly where soil moisture
contents were high. The average surface displacement
ranged from 2 mm to 7 mm, depending on site and season.
At all sites soil freezing over this short time scale was
restricted to the surface layers. Few cycles penetrated
to 10 cm, and none to 20 cm.

   James (1971) also reports frost heaving by needle
ice during diurnal freeze-thaw cycles in March 1970 at
Nottingham, England. Frost penetration was around 3 mm.
However, during a ten day period of soil freezing in
February 1970 frost heave of some 13 mm accompanied frost
penetration to below 10 cm. The monitored soil consisted
of gravelly silt.

   In terms of periglacial mass movement processes it may

75

be concluded that the annual freezing and thawing cycle produces significant ice segregation where site conditions are suitable and is of major significance to slope stability and soil displacement. Diurnal cycles are often restricted by snow cover, occur in the surface layers only, and often produce needle ice rather than ice lenses. Although needle ice may cause downslope displacement of individual soil grains, diurnal freezing and thawing is not considered to be important in downslope soil movement below two or three centimetres depth. Such a conclusion is supported by Washburn's detailed measurements of soil movements in the Mesters Vig District, N.E. Greenland (Washburn 1967).

Frost creep as a component of periglacial mass movement

As we have seen in the introductory part of this chapter, where frost heaving occurs on an inclined plane, resettlement during thaw is associated with downslope displacement of the heaved soil mass, due to gravity. If data are available on the amount of frost heave at a given site, the maximum potential frost creep generated by vertical resettlement following one freeze-thaw cycle may be simply computed from:

$$C = h \tan\theta \qquad (3,15)$$

where C = downslope displacement at soil surface
       h = heave normal to the slope at the soil surface
       θ = surface slope

This assumes that the soil resettles to its initial level following heave. Davison (1889) and Washburn (1967) have shown that vertical resettling is unlikely, so the actual creep value is likely to be less, according to Davison about one-half that predicted by equation 3,15. Due to the variability of soil moisture contents during thaw, and the variable amount and effectiveness of soil colloids from one site to another, it is impossible to predict precisely the downslope displacement of the soil due to frost heave and resettlement. However, equation 3,15 provides a means of qualitatively assessing the relative importance of frost creep where field measurements of frost heave and total downslope soil movements are available.

Field studies of frost creep

Among the most detailed and meticulous studies of periglacial mass movements is that by Washburn, in the Mesters Vig District of N.E. Greenland (Washburn 1967). Here the mean annual temperature between 1952 and 1961 was $-9.7^{\circ}C$, with absolute maximum $21.0^{\circ}C$, and minimum $-44.2^{\circ}C$. Slope movement data were obtained by repeated theodolite survey of target cones attached to pegs buried to different depths in the soil. All displacements of the targets were recorded relative to the vertical plane

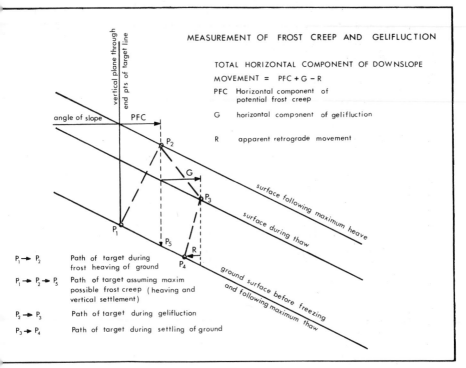

Figure 43 Frost creep, gelifluction and retrograde components of target movement as related to the vertical plane through the end points of a target line (Washburn 1967).

through the fixed end points of the target line. Washburn defined four categories of target movement; jump, the movement recorded between the last target reading of one year and the first reading of the next, resulting mainly from frost heave (figure 43); gelifluction, the movement due to flow; retrograde movement, backward movement upslope of the vertical plane due to the tendency of the soil to resettle normal to the slope; and September movement, caused either by soil freezing and consequent frost heaving, or gelifluction. Since little thawing occurs at this time of year, September movement is probably mainly due to frost heave.

Inspection of figure 43 shows that the jump value minus the retrograde movement value gives the horizontal component of downslope movement due to creep, the total downslope movement being the sum of the creep and gelifluction. In order to estimate the frost creep component of downslope movement in the field, measurement of the position of the target P_2 in figure 43 must be made before surface thaw begins so that no gelifluction component is added to the potential frost creep value.

Table 9 Frost creep and potential frost creep, Mesters Vig, Greenland (Washburn 1967).

Target line	Slope	Soil	Average estimated frost creep as % potential frost creep 1956-61	Average estimated frost creep as % potential frost creep control periods		Average estimated frost creep, per year 1956-61 mm	Average estimated frost creep, per year control period mm		Ratio estimated frost creep to measured gelifluction 1956-61	Ratio estimated frost creep to measured gelifluction Control Period	
				1959-60	1960-61		1959-60	1960-61		1959-60	1960-61
ES7 (dry)	10°-14°	Silty diamicton soil	50%	70%	70%	4.4	4.0	6.0	2.0 : 1	1.5 : 1	5.3 : 1
ES7 (wet)	10°-14°	Silty diamicton soil	90%	100%	70%	20.2	18.0	13.0	1.6 : 1	1.0 : 1	0.5 : 1
ES6 (wet)	2.5°-3°	Silty, sandy clay diamicton soil	cf.40%	no data	no data	4.8	no data	no data	1.7 : 1	no data	no data
ES8 (dry)	11°-14°	Silty diamicton soil	70%-80%	no data	70%	18.6		9	2.3 : 1	no data	0.36 : 1
ES8 (wet)	11°-14°	Silty diamicton soil	70%-90%	no data	70%	26.2		16	2.8 : 1	no data	0.71 : 1

By measuring the position P_2 of targets as they emerged in a still frozen state from beneath a covering of snow Washburn was able to estimate the creep component of downslope soil movement in the Mesters Vig study. At a target line with a silty diamicton soil and 10^0 to 14^0 slope (line ES 7, Washburn 1967) Washburn found that in areas subject to dessication in summer ('dry sites') over the years 1959-60 and 1960-61 actual frost creep was on average 70% of the potential frost creep (h tan θ, equation 3,15), while at 'wet' sites which remained saturated during the summer, actual frost creep averaged 90% of the potential value. Over the longer period of five years the actual creep recorded in the dry sites averaged 50% of the potential value and in the wet sites was again 90% of the potential. These long term values are considered to be maxima, since they include small amounts of gelifluction in some of the jump measurements, due to survey of target positions P_2 after slight surface thawing had begun. They are regarded however as the best guide to the maximum actual creep over the longer term.

At the dry sites the average amount of frost creep over the five year period 1956-1961 was in total 22 mm at the surface, giving an average annual rate of 4.4 mm. In the wet sites the averaged frost creep over the five years was 101 mm, giving an annual average of 20.2 mm. In the dry sites frost creep was found to be more important than gelifluction, while in the wet sites the contributions of creep and flow were more nearly alike.

Similar estimations of frost creep in relation to potential frost creep were made at two other target lines in Washburn's study, and are presented, together with the data above in table 9.

It is clear that the amount of frost creep and the amount of gelifluction varied from site to site and from year to year, depending on such factors as soil moisture content during freeze, moisture supply from melting ice lenses, snow melt and rainfall during summer, depth of summer thaw etc.

Benedict (1970) measured soil displacements in an alpine environment on the Niwot Ridge, Colorado Front Range (mean annual temperature 1953-64, -3.3^0C, absolute maximum and minimum, 19.4^0C and -36.7^0C). Lines of pegs were inserted across the slope to depths of 25 cm and 50 cm, and their displacement measured against a tensioned steel wire strung between benchmarks at either end of the peg line. Results are summarised in table 10.

Again creep was found to be highly variable both temporally and spatially, and although no specific measurements were made, Benedict concluded that solifluction was dominant in the wetter sites, and creep was dominant in the drier sites.

Table 10   Frost creep measurements, Niwot Ridge, Colorado
Front Range (Benedict 1970)

Feature	Elevation m	Slope	%silt/clay in fraction finer than 2.0 mm	Average annual frost creep mm/yr	Period
Turf-banked lobe No. 499	3,640	13°	40-50	max. potential creep 30 mm, but actual creep much less.	1965-66
Turf-banked lobe No.45	3,480	6°-7°	30-40	negligible	1965-66
Turf-banked terrace No. 19	3,440	11°-12°	30-45	0-3.2	1964-67
Turf-banked lobe No. 26	3,480	13°-14°	30-55	up to 7.2	1962-67
Stone Streams 502 and 503		12°-13°	40-55	0.4-6.0 av. 3.7	1965-67
Between stone streams 502 & 503		16°-18°	35	av. 1.2	1965-67

Harris (1972) in his study of mass movements in turf-banked solifluction lobes on a till slope in the Okstindan Mountains of North Norway (mean annual temperature cf. -2.5°C) attempted to evaluate the relationship between frost creep and gelifluction. Potential frost creep was calculated from frost heave measurements by substituting in equation 3,15 above, and compared with measured rates of soil movement, as shown in table 11.

The ratios of potential frost creep to measured soil movement at the surface range from 0.03:1 to 0.48:1, and bearing in mind Washburn's data on the relationship between potential frost creep and actual frost creep (table 9) it is apparent that frost creep at this site in Norway was less than gelifluction. Jahn (1961) reports potential frost creep of 10 mm on a 4° slope in Spitzbergen compared with measured displacements of 30 mm, and similarly concludes that the soil movements observed were largely due to flow rather than creep.

Table 11 Potential frost creep, Okstindan, North Norway, 1969-71

Location	Slope angle (degrees)	Summer resettlement perpendicular to the slope, following frost heave (mm)		Potential frost creep (mm)		Observed soil movement (mm)	
		1970	1971	1969-70	1970-71	1969-70	1970-71
Site A	5°	35.0	88.0	3.06	7.69	17.5	
Site B	15°	18.0	21.0	4.80	5.6	10.0	0-48.0
Site C	13°	8.0	?	1.8	?	(1)60.0[+] (2)40.0[*]	8.0-26.0

+ slope 11°

* slope 6°

Williams (1960) measured downslope soil movements at Schefferville (mean annual temperature -5°C to -6°C) using buried flexible tubes and a specially constructed probe. On a slope of 4° to 9° formed of a silty frost susceptible soil, downslope movements of 93.7 mm per annum were recorded, movement decreasing with depth and dying out at 0.65 m below the surface. At this site Williams compared the amount of frost heave required in each successive 50 mm thick layer of soil to give potential frost creep equal to the increment of movement recorded for that layer. It was found that for individual 50 mm layers frost heaves of more than 100 mm were required in some cases, and such enormous heaves were considered impossible. The total heave at the surface necessary to produce the recorded soil displacements amounted to some 66.7 cm, while in fact tubes were only heaved by 5 to 10 cm. Williams therefore concluded that frost creep was unlikely to have played a major role in the mass movement recorded.

The field studies mentioned above illustrate the difficulty of separating downslope soil movement due to frost creep and that due to other processes, noteably gelifluction. The general concensus appears to be that frost creep is of only minor importance compared with gelifluction in most fine grained soils (McRoberts and Morgenstern, 1974), but becomes relatively more important in drier sites. Since frost heave of the soil during freeze, and subsequent consolidation of the soil during thaw are vital elements of both frost creep and gelifluction, and both processes operate simultaneously on most periglacial slopes where soils are frost susceptible, it is probably best to consider their combined effects, under the umbrella term solifluction (chapter 1).

Talus creep

Talus creep is defined by Sharpe (1938) as the slow downslope movement of a talus or scree or any of the material of a talus or scree, and he considers such movements to be most rapid in cold regions where freezing and thawing of ice in the interstices of the rock waste leads to expansive forces within the surficial material. Such expansive forces are not generated by ice lens growth in such coarse material, but must result from the expansion associated with phase change of water at the points of contact between rock fragments. Beskow (1935) suggested a mechanism of needle-ice crystal growth between adjacent particles of unsaturated sand to explain 'heaving' and such a mechanism might operate in talus. If 'heaving' of the surficial layers occurs due to freezing, resettlement during thaw is likely to be vertical due to the large mass of the individual rock fragments in relation to any apparent cohesion between grains due to wetting films of water (figure 26).

Other causes of talus creep include the washing out of interstitial fines (Sharpe 1938) and settling due to consolidation of the talus mass (Rapp 1960).

Talus creep has been measured as part of a large scale study of slope processes in the Kårkevagge region of northern Scandinavia (Rapp 1960). The mean annual temperature here is around -1.5°C (Riksgränsen 1901-30). Diurnal frost cycles are weak, but more frequent in spring than in autumn.

Rapp monitored talus movement by means of resurvey of stakes inserted 40-50 cm into the talus in lines running downslope, smaller stakes inserted to depths of 15-20 cm, and painted boulders and cobbles. The position of stakes and boulders was checked with a steel tape once every summer. In addition to talus creep, movements were noted due to consolidation, particularly near the top of the talus where the supply of fresh debris was greatest, small slides, particularly near the base of the talus where they formed tongues of debris, and individual sliding and rolling of grains.

On an active talus cone with slope angles up to 38° the recorded movements of the stakes and boulders embedded in the surface indicated a fairly continuous creep movement of about 10 cm per year in the higher part of the

slope decreasing to zero at the base. Talus creep was
considered to be relatively rapid due to the schistose
rock fragments of which the slope was composed. At any
particular point movement varied by as much as a factor
of four between successive years. The thickness of the
moving layer was considered to be 10-20 cm. Sliding and
rolling of boulders over the surface was revealed by
marked boulders which moved as much as 10 m in two years.
Rapp suggests that the lack of talus creep and surficial
boulder sliding at the base of the talus slopes indicates
that these processes redistribute material from the upper
to the lower parts of the slope, where slides, gullying
and mudflows become the chief agents of removal.

Contemporary movement processes on low angle scree
slopes in the Lake District of Northern England were
monitored through the winter of 1961-62 by Caine (Caine
1963). Here coarse blocks form a layer up to 60 cm thick
above a finer subsoil. Movement was measured by resurvey
of painted stones and by the deformation of flexible
plastic tubes inserted to a depth of approximately 40 cm.
On slopes of around 15^0 surface movement rates were
between 15 and 20 cm, but the plastic tubes, inserted
below the stony surface layer indicated that movement
immediately below this layer was between 2.5 cm and 11 cm.

Caine concluded that the surface stones move
independently of the underlying soil by sliding and
rolling, particularly when the ground is saturated. Such
conditions occur particularly when thawing is associated
with heavy rainfall. The subsurface material moves by
frost induced creep. Evidence for ground heaving came
from the ejection of the plastic tubes by frost heave.
Five periods of ground freezing were observed through the
winter of 1961-62 producing mean frost heaving values of
2.2 cm, 3.8 cm, 2.5 cm, 9.5 cm and 7 cm. The last two
observations were considered 'very doubtful' however. The
rate of movement fell rapidly with depth, and died out at
about 10 cm below the surface. The profile of movement
was concave, probably because the incidence of freeze-thaw
cycles decreased with depth.

With data for only one year it is clearly impossible
to give long-term averages of talus creep in these Lake
District screes, but it would appear unlikely that rates
of movement exceed those recorded by Rapp in Scandinavia.

Needle-ice induced frost creep

In a series of laboratory experiments aimed at
simulating solifluction processes, Higashi and Corte (1970)
measured soil movements in an inclined cabinet subject to
freezing and thawing. The soil used was a very frost
susceptible silty clay. Needle ice was observed to lift
surface particles perpendicular to the slope, and as thaw
commenced bending of the needle and the heave-resettlement
displacement were proportional to the slope angle, so that

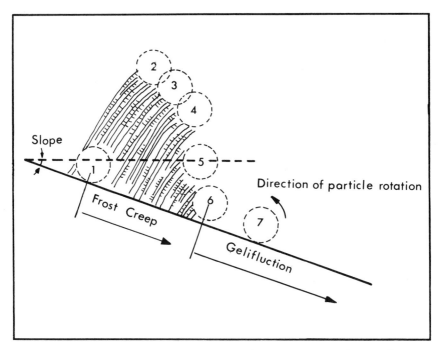

Figure 44   Schematic diagram of frost creep of a soil
             particle due to growth and decay of needle ice
             (Higashi and Corte 1971).

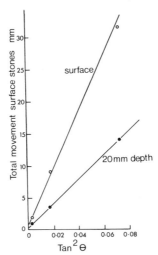

Figure 45   Total movement of stones at the surface and at
             20 mm depth after three cycles of freeze-thaw
             as related to angle of slope (Higashi and Corte
             1971).

their combined effect on the creep of surface material resulted in surface movement proportional to $\tan^2\theta$ rather than $\tan\theta$, where θ was the angle of slope (figure 45). This relationship held for surface creep and creep at a depth of 20 mm.

The work of needle ice is discussed by Washburn (1973), who notes that it can lift stones and cobbles as well as finer soil material, and may be a significant factor in frost sorting and frost creep. Needle ice is the result of diurnal ice segregation near the surface, resulting from upward migration of water to a stationary freezing plane (Outcalt 1969, 1970, 1973). Needle ice requires a soil containing sufficient fine material (frost susceptible, see for instance Gradwell 1954), sufficient water, and with a sufficiently high permeability to allow rapid moisture migration, preventing downward movement of the freezing plane (Washburn 1969). Needle ice is widely reported from both arctic and alpine regions, and indeed temperate regions such as England (e.g. James 1971), and must therefore be considered as a potential cause of downslope creep in the near-surface soil layers.

Instrumentation for frost creep measurement

Under periglacial conditions soil movement due to frost creep has been recorded mainly by annual measurement of surface markers, pegs, or buried tubes. More sophisticated instrumentation has however been developed, particularly for work in temperate environments where precision is essential. A useful survey of instrumentation for measuring soil creep is provided by Anderson and Finlayson (1975), and many of the techniques described by these authors could be adopted for use in periglacial regions.

4. PERIGLACIAL SLOPE STABILITY ANALYSIS

Introduction

Since Andersson's account of mass wasting in Bear Island it has become generally accepted that mass movement processes are particularly active in periglacial regions, although early workers sometimes exaggerated their rates of operation (e.g. Högbom 1914). Among the most commonly cited reasons for rapid mass wasting include soil freezing and thawing leading to accelerated rates of soil creep, and high water content in the active layer resulting from: (a) snow melt and the thawing of segregation ice during spring and summer; (b) impedence of drainage by permafrost or seasonally frozen ground; and (c) low evaporation rates. The response of a given slope to such environmental conditions depends largely on its engineering properties and geometry. Soil strength parameters and their relation to water content are examined below, with special reference to thawing soils.

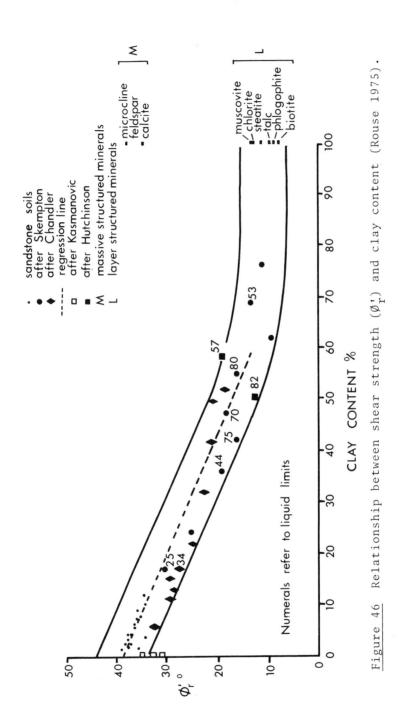

Figure 46   Relationship between shear strength (ϕ'_r) and clay content (Rouse 1975).

Geotechnical strength parameters

Shear strength

Parameters of soil shear strength may be determined using either drained or undrained tests in the shear box or the triaxial apparatus.

In the drained test, samples are first consolidated under the applied confining stress and during testing water is allowed to drain from the saturated specimen so that no pore water pressures are developed, $u = 0$, and the effective stresses are equal to the applied stresses. This not only requires time for consolidation prior to shearing, but the application of a deviator stress at such a low rate of strain that any pore pressures induced in the sample during shearing have time to dissipate.

More rapid shearing is possible using the undrained test, where following consolidation, further drainage of the specimen is prevented, and pore pressures developed during shear are measured. These may then be subtracted from the total deviator stress in the triaxial test, or the normal load in the shear box test in order to find c' and $\emptyset'$. In the triaxial test these parameters are obtained from Mohr's circle stress and rupture diagrams (Terzaghi and Peck 1967, p. 100-106), and in the shear box test from the relationship between stress at failure and effective normal load.

Following shearing, soil strength is less due to reduction in both cohesion and internal friction. Disruption of the colloidal bonds across the shear plane reduces cohesion, often to zero, while reorientation of soil grains adjacent to the shear plane reduces their interlocking and lowers the friction along the shear surface. Residual cohesion (c_r') and friction ($\emptyset_r'$) values must therefore be used in the analysis of slopes where failures and displacements have already taken place, as is commonly the case in periglacial areas. Skempton (1964) and Rouse (1975) have shown a general qualitative relationship between texture and residual shear strength, $\emptyset_r'$ tending to decrease as clay content increases (figure 46).

Generally soils in the present day periglacial zone tend to be silts or sands rather than clays and are therefore often frictional non-cohesive soils with $\emptyset_r'$ values ranging from $20°$ to $35°$ or more. During the Pleistocene, however, clay soils such as the London Clay, Weald Clay, Gault Clay and Liassic Clay in England were subject to periglacial mass movements (table 12).

Index properties

After a cohesive soil has been remoulded its consistency can be changed at will by increasing or

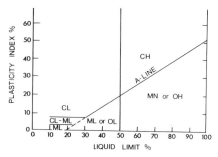

Figure 47   Unified soil classification system, U.S.B.R.
             (1963).

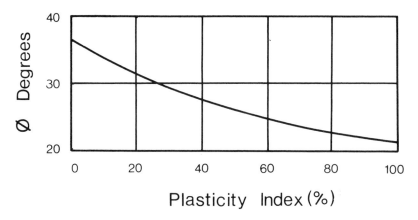

Figure 48   Relationship between φ and Plasticity Index
             for clays of moderate to low sensitivity
             (Terzaghi and Peck 1967).

decreasing the moisture content. If the moisture content
of a clay slurry is gradually reduced the clay passes
from as liquid state, through a plastic state and finally
into a solid state. The transitions from liquid to
plastic and from plastic to solid were defined originally
by Atterberg (1911), and the boundaries between the
states of consistency are called Atterberg Limits.
Standard laboratory procedures are available for their
measurement (British Standards 1377, 1975). The
numerical difference between the liquid limit and the
plastic limit is called the plasticity index, and
represents the range of moisture contents over which the
soil is plastic. Casegrande (1932) showed that the
plasticity index is directly related to liquid limit and
developed a classification of soils according to their
liquid limits and plasticity indices. This
classification has been modified by the U.S. Bureau of
Reclamation (USBR 1963) to produce the Unified Soil
Classification System. Fine-grained soils are divided

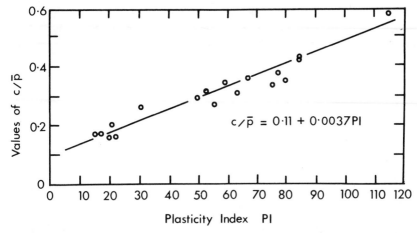

$$c/\bar{p} = 0{\cdot}11 + 0{\cdot}0037 PI$$

Figure 49   Relationship between $C_u/\bar{p}$ and Plasticity Index (Skempton 1957).

into inorganic clays (C, figure 47), inorganic silts (M) and organic soils (O). The soils are further divided into those having liquid limits higher than 50% (H) and those having liquid limits lower than 50% (L).

Various attempts have been made to relate Atterberg limits to shear strength parameters. Terzaghi and Peck (1967) for instance note that the value of ϕ is generally related to the plasticity index (figure 48) but they state that exceptional cases do occur, and that Bjerrum and Simons (1960) observed a scattering from the curve of the order of 5^0. From our knowledge of the relationship between soil texture and strength parameters we can say that in general sandy and silty non-cohesive soil have low plasticity, while cohesive clays show higher plasticity index values.

Skempton (1957) has illustrated the relationship between plasticity index and the cohesive strength of normally loaded soils by considering their undrained shear strengths where pore pressures during shear reduce ϕ' to zero. If $\bar{p}$ is the effective overburden pressure at the depths corresponding to the strength tests and c_u is the undrained shear strength when $\phi' = $ zero, figure 49 illustrates the relationship between $c_u/\bar{p}$ and the plasticity index. As the plasticity index increases so the cohesive strength of the soil increases.

In a detailed study of index properties of sediments subject to periglacial processes at Keewatin, N.W.T. Canada, Shilts (1974) has investigated the relationships between texture and liquid limits. His results are summarised in figure 50. It is apparent that progressively finer grain sizes, as expressed by both phi mean grain size and percent clay are associated with progressively higher liquid limits, although the degree of

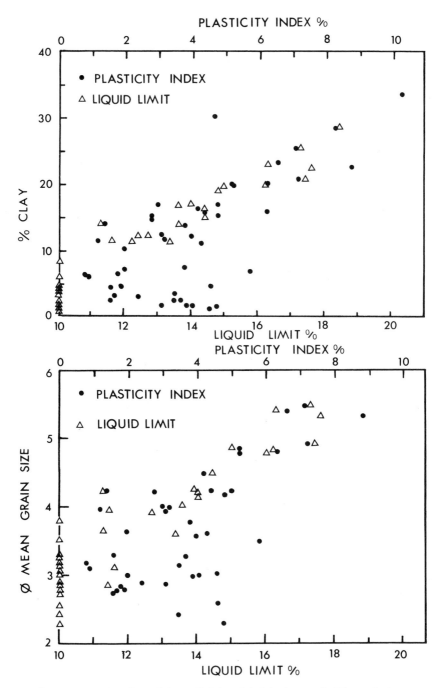

Figure 50   Relationship of Liquid Limit and Plasticity
Index with clay content and phi mean grain
size, Keewatin, N.W.T., Canada (Schilts 1974).

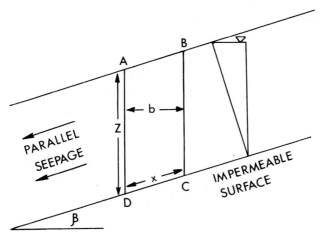

Figure 51   Slope model for infinite slope analysis of
A.W. Skempton and F.A. DeLory (1957).

scatter is large. Increasing values of the liquid limit
are also associated with increasing plasticity index.

The liquidity index is a measure of the closeness of
a soil's water content to its liquid limit, and is defined
as:

$$LI = \frac{w - PL}{LL - PL}$$

where w = water content as % of dry weight. If the water
content exceeds the liquid limit, LI>1.0, and remoulding
transforms the soil into a thick viscous slurry. If the
natural water content is less than the plastic limit (LI
is negative) the soil cannot be remoulded.

Some examples of the engineering properties of sediments
subjected to periglacial mass-wasting are given in table
12.

Conventional slope stability analysis

In the analysis of periglacial mass movements in
unfrozen unconsolidated sediments it may be assumed that a
frozen subsoil prevents deep seated slides of a rotational
form. Hughes (1972), Issacs and Code (1972) and McRoberts
and Morgenstern (1974b) have however described deep seated
slides through frozen ground, but it appears that the base
of these failures generally passes through underlying
unfrozen clays. Such landslides will not be considered
here.

Conventional analysis of the soil mechanics of slope
failures assumes failure along a slip surface, which in
the case of a thawing active layer may be assumed to lie
immediately above the interface between frozen and

Table 12 Engineering properties of sediments subjected to periglacial mass-wasting.

MATERIAL	LOCATION	φ' degrees	φ'r degrees	c' kN/m²	c'r kN/m²	LL %	PL %	PI %	AUTHOR
Clayey silty gravelly sand	Mesters Vig, Greenland					19	15	4	Washburn (1967)
sandy silty clay						33	22	11	
Sandy silt	Taseriaq, Greenland					14 15 22	10 13 16	4 2 6	Everett (1967)
Illitic clay	Mackenzie Valley, Canada		23		0				McRoberts & Morgenstern (1974)
Sandy clayey silt	Vest Spitsbergen		36		0				Chandler (1972)
Silty sand	S.W. Spitsbergen					17-20			Jania (1977)

Table 12 (continued)

MATERIAL	LOCATION	φ' degrees	φ'r degrees	c' kN/m²	c'r kN/m²	LL %	PL %	PI %	AUTHOR
Silty fine sand	Okstindan Mts., N. Norway	35	29-33	0	0	21-32 av.25	17-27 av.21	2-9 av.4	Harris (1977)
Silty sand	Banks Island Canadian Arctic					17-32 av.23	10-23 av.15	7-9 av.8	French (1974)
Stony silty sand head (3-14% clay)	S.Devon England					21-29	17-19	4-11	Mottershead (1971)
Stony sandy silt head	Beddau, S.Wales Coalfield					26	16	10	Harris & Wright (1980)
Clayey sand, sandy clay head	Northamptonshire, England	23			0	33-61	18-32	4-37	Chandler (1970a)
Clay head	Rutland England	18.5			0	71-76	29-32	42-44	Chandler (1970b)

Table 12 (continued)

MATERIAL	LOCATION	φ' degrees	φ'r degrees	c' kN/m²	c'r kN/m²	LL %	PL %	PI %	AUTHOR
Clayey head (Fuller's earth parent material)	Bath, England	13.5		2		47	19	28	Chandler et al. (1976)
	Bath, England	13.5		2		77	25	52	
Silty sand head	Bath, England					44	19	25	
Lias clay head	Bath, England	12.5		1.5		48	23	25	
Clay head	Sevenoaks S.E. England	14			1	42	24	18	Skempton & Weeks (1976)
		16			0.2	52	26	26	

unfrozen soil. Although Burt and Williams (1976) have
shown that frozen soil may not be totally impermeable,
permeabilities are several orders of magnitude less than
equivalent unfrozen permeabilities, so that for practical
purposes the frozen soil may be considered impermeable.
Seepage of ground water must therefore take place
laterally downslope.

The infinite slope analysis of Skempton and DeLory
(1957) provides a suitable model of the slope conditions
described above. Figure 51 shows the assumptions made in
the analysis, which include a planar slip surface parallel
to the ground surface, an infinite slope (in other words,
ignoring boundary conditions), water table at the surface,
seepage parallel to the surface, and a uniform soil.
Consider a small element of soil in ABCD in figure 51;

$$\text{shear stress on the slip surface, } \tau = \frac{W\sin\beta}{x} \quad (4,1)$$

but $W = Zb\dot{\gamma}$, where γ = unit weight of saturated $\quad (4,2)$
soil, and $x = \dfrac{b}{\cos\beta}$

therefore $\tau = \gamma Z\cos\beta\sin\beta$ $\qquad\qquad (4,3)$
Normal effective stress on the slip surface, $\sigma'_n = \quad (4,4)$
$\dfrac{W\cos\beta}{x} - u$, substituting (4,1) and (4,2),

$$\sigma'_n = \dot{\gamma}Z\cos^2\beta - u \qquad\qquad (4,5)$$

with water table at the surface and parallel seepage,
$\qquad u = \gamma_w h$, where $\dot{\gamma}_w$ = unit weight of water $\quad (4,6)$
$$u = \gamma_w Z\cos^2\beta \qquad\qquad (4,7)$$
therefore,
$$\sigma'_n = \gamma Z\cos^2\beta - \dot{\gamma}_w Z\cos^2\beta \qquad (4,8)$$
substituting in the Coulomb equation,
$$s = c' + (\gamma Z\cos^2\beta - \gamma_w Z\cos^2\beta)\tan\phi' \quad (4,9)$$
where s = strength of soil.
Taking a factor of safety $F = \dfrac{s}{\tau}$, and substituting (4,2)
and (4,8)
$$F = \frac{c' + (\gamma Z\cos^2\beta - \gamma_w Z\cos^2\beta)\tan\phi'}{\gamma Z\cos\beta\sin\beta} \quad (4,10)$$

$$\text{and if } c' = 0, \quad F = \left(\frac{\gamma - \gamma_w}{\gamma}\right)\frac{\tan\phi'}{\tan\beta} \quad (4,11)$$

When $F = 1$ the slope is at its limiting state, and
$$\tan\beta = \frac{(\gamma - \gamma_w)}{\gamma}\tan\phi' \qquad (4,12)$$

Since as a general rule γ is approximately equal to $2\gamma_w$,
$$\tan\beta = \text{approximately } \tfrac{1}{2}\tan\phi' \qquad (4,13)$$

In the analysis of both active and fossil periglacial
slopes it is generally appropriate to use residual
parameters c'_r and ϕ'_r.
The magnitude of the total pore pressure can be expressed
in terms of the ratio:

$$r_u = \frac{u}{\gamma Z} , \text{ where } \gamma Z \text{ is the total stress} \qquad (4,14)$$

substituting in equation (4,10) we have,

$$F = \frac{c' + \gamma Z(\cos^2\beta - r_u)\tan\phi'}{\gamma Z\sin\beta\cos\beta}, \text{ which may be written as:}$$

$$\sin\beta\cos\beta = \frac{1}{F}\left[\frac{c'}{\gamma Z} + (\cos^2\beta - r_u)\tan\phi'\right] \qquad (4,15)$$

and if c' = 0

$$\sin\beta\cos\beta = \frac{1}{F}(\cos^2\beta - r_u)\tan\phi' \qquad (4,16)$$

For hydrostatic conditions r_u = approximately 0.50 (since γ_w = approximately $\frac{1}{2}\gamma$), and higher values give a measure of excess pore pressures. Where the slope angle β is less than the limiting slope angle under hydrostatic conditions substitution in equations 4,15 or 4,16 with F = 1 enables calculation of the minimum value of r_u to initiate movement.

This type of stability analysis has been applied to active periglacial mass movements (e.g. Chandler 1972; Harris 1972, 1977; McRoberts and Morgenstern 1974) and to fossil Pleistocene slope deposits (Weeks 1969; Chandler 1970; Skempton and Weeks 1976). The results of these analyses are summarised in table 13.

It is apparent from table 13 that very high pore water pressures are necessary in all the sites analysed in order to initiate movement. It appears that such high pore pressures are widely developed under periglacial conditions.

Thaw consolidation theory

A mechanism whereby pore pressures may develop has been described by Morgenstern and Nixon (1971), Nixon and McRoberts (1973), Nixon and Morgenstern (1973), McRoberts and Morgenstern (1974), Nixon and Ladanyi (1978) and McRoberts (1978), and is referred to as the thaw consolidation theory. As pointed out by McRoberts (1978), the concept of thaw consolidation was outlined by Taber (1943) who described the loss in strength of ice-rich soils subjected to rapid thawing and slow drainage. Under these conditions water is released by the melting of ice lenses at a faster rate than it can be expelled by the soil, so that in the process of consolidation some of the load is transferred to the pore water with resulting loss in strength.

We have seen in chapter 3 that the freezing of fine grained soils leads to ice segregation and water being drawn to the freezing plane to feed the growing ice lenses. The development of ice lenses in a freezing soil has two important consequences with regard to its stability during thaw. Firstly the thickness of the frozen soil is increased by an amount defined by the heaving ratio, so that the density of the frozen soil is

Table 13 Slope stability analyses of periglacial slopes.

Author	Area	Soil type (failure type)	Residual strength c'_r (kN/m²)	strength ϕ'_r (degrees)	Predicted angle (degrees)	Failed angle (degrees)	r_u for failure
Chandler (1972)	Vestspitsbergen	Sandy clayey silt (mudslide)	0	36	20	6-12	0.84 to 0.72
Harris (1977)	Okstindan, Norway	Fine silty sand (solifluction)	0	29	15-16	5-14	0.84 to 0.52
McRoberts & Morgenstern (1974a)	Mackenzie Valley	Illitic clay (skinflow and bimodalflow)	0	23	12.5	3-9	0.87 to 0.61
Weeks (1969)	Kent, England	Disturbed Weald Gault & London clay (slab slide)	0-1.38	12.4-15.5	6.8-8.1	3-7	0.75 to 0.50
Chandler (1970)	Northamptonshire, England	Sandy clayey silt (slab slide)	0	16	8.8	4	0.75
Skempton & Weeks (1976)	Kent, England	Disturbed Weald clay (slab slide)	0.2 / 0	16 / 18	8	0.6-3	0.97 to 0.83

less than its density when unfrozen. Thawing is therefore associated with settlement of the soil grains and an increase in bulk density. Secondly, since water is drawn to the freezing plane to supply the growth of ice lenses, the soil in its frozen state may contain a considerable volume of water in excess of its unfrozen pore space. This excess water content of the frozen soil is given by equation 3,2 as;

$$W_{ex} = \gamma_i H \qquad \text{where } W_{ex} = \text{the excess water content}$$
$$\gamma_i = \text{the density of ice}$$
$$\text{and} \quad H = \text{the heaving ratio}$$

Thawing must therefore be associated with the expulsion of the excess water as voids formerly occupied by ice lenses are closed by settlement.

However, the drainage conditions similar to those described in the conventional stability analysis model (figure 51) must apply, that is with lateral seepage over an impermeable frozen sub-surface, either permafrost, or a still-frozen layer where seasonally frozen ground is thawing from the surface. The rate at which water can seep away is limited by the permeability of the soil so that a finite time is required for the excess water released by melting ice lenses to escape. If thawing is slow, water can flow from the soil as fast as it is produced, and no excess pore pressures will develop. However, rapid thawing may release water faster than it can flow away leading to the generation of excess pore pressures, and in consequence, a reduction in the frictional strength of the soil.

Morgenstern and Nixon (1971) developed a one-dimensional thaw consolidation theory based on Terzaghi's theory of consolidation and the Neuman solution for the penetration of a thaw plane through a frozen soil (chapter 2). A detailed discussion of thaw consolidation is given by Nixon and Ladanyi (1978), including the one-dimensional thaw consolidation theory.

When a saturated soil is loaded it suffers consolidation. This consolidation results from a decrease in the void ratio e, and consequent expulsion of water. The coefficient of volume compressibility $m_v (m^2/kg)$ is defined by Terzaghi and Peck (1967) as the compression of the soil per unit of original thickness due to a unit increase of pressure. If x is the thickness of the soil under pressure p an increase in pressure from p to p + Δp reduces the thickness x by

$$\Delta x = x.\Delta p.m_v \qquad (4,17)$$

Consolidation tests have shown that for equally thick layers of different soils the time required to reach a given degree of consolidation increases in direct proportion to m_v/k, where k is the coefficient of permeability. The ratio,

$$c_v (m^2/sec) = \frac{k}{m_v} \cdot \frac{1}{\gamma_w} \qquad (4,18)$$

is known as the coefficient of consolidation, and governs

the rate at which excess pore water can be squeezed out of
the thawed soil. It can be shown that the excess pore
pressure u generated by consolidation and the need to
expel water is given, with respect to time as;

$$\frac{du}{dt} = c_v \frac{d^2u}{dx^2}$$ (4,19)

(see for instance Terzaghi and Peck 1967 p. 178), where
in a thawing soil x is the depth measured from the ground
surface.

In a thawing soil it is assumed that consolidation
results from the expulsion of excess water released by
the melting of segregation ice, so that the rate of
movement of the thaw plane influences consolidation by
controlling the release of excess water, and hence the
pressure gradient in the thawed soil, and by increasing
the thickness X of the thawed soil layer. The movement
of the thaw plane is given by equation 2,19 as;

$$X = \alpha t^{\frac{1}{2}}$$

where X is the depth from the surface.
The two time dependent factors, rate of thaw and rate of
consolidation as defined by c_v are combined in the thaw-
consolidation ratio R, where

$$R = \frac{\alpha}{2\sqrt{c_v}}$$ (4,20)

Hence if c_v remains constant, the faster the rate of
thawing as expressed by the downward movement of the thaw
plane, the higher the value of R. Morgenstern and Nixon
show that for an infinite soil mass thaw-consolidating
under self-weight (that is with no applied loading other
than overburden pressure), the excess pore pressure is

$$u = \frac{\gamma'X}{1 + \frac{1}{2R^2}}$$ (4,21)

where $\gamma' = (\gamma - \gamma_w)$ and is the submerged unit weight of the
soil. $\gamma'X$ is the effective stress after complete
dissipation of excess pore pressures. Nixon and McRoberts
(1973) use the Stephan solution to the Neuman equation
(chapter 2, equation 2,10), to define α, as shown in
figure 22, where

$$x = \sqrt{\frac{2k_uT_st}{L}} \quad \text{and} \quad \alpha = \sqrt{\frac{2k_uT_s}{L}}$$

where k_u = unfrozen thermal conductivity, T_s = step
temperature change at the surface, L = latent heat of soil
and t = time.

The value c_v may be obtained from laboratory
measurement of k and m_v, the saturated permeability, and
the coefficient of volume compressibility respectively,
and substitution in equation 4,18.

Morgenstern and Nixon (1971) provide a graph of
normalized values of excess pore pressures computed from
equation 4,21, where pore pressures in excess of
hydrostatic are expressed as $u/\gamma'X$ (figure 52). Taking

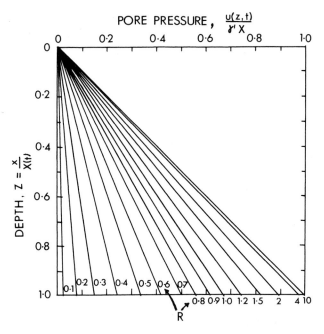

Figure 52 Excess pore pressures due to thaw consolidation with no applied load (Morgenstern and Nixon 1971).

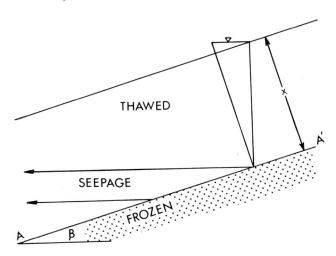

Figure 53 Infinite slope analysis, model of thawing slope suffering thaw consolidation (McRoberts and Morgenstern 1974).

$\gamma = 2\gamma_w$ (e.g. Chandler 1976), we may express the excess pore pressures in terms of r_u, where $r_u = 0.5 + \frac{1}{2} \cdot \frac{u}{\gamma'X}$,

Table 14 Estimated values of R for periglacial slope failure

Author	Area	r_u for failure	Estimated R
Chandler (1972)	Vestspitsbergen	0.72 to 0.84	0.64 to 1.0
Harris (1977)	Okstindan, Norway	0.52 to 0.84	0.15 to 1.0
McRoberts & Morgenstern (1974)	Mackenzie Valley	0.61 to 0.87	0.38 to 1.2
Weeks (1969)	Kent, England	0.54 to 0.75	0.2 to 0.7
Chandler (1970)	Northamptonshire, England	0.75	0.7
Skempton & Weeks (1976)	Kent, England	0.89 to 0.97	1.4 to 3.0

since under hydrostatic conditions, with the water table at the surface, $r_u = 0.5$. From figure 52 it can be seen that the excess pore pressure generated at the thaw plane increases as R increases, as would be anticipated from the definition of R (equation 4,20). Using figure 52 and knowing the r_u necessary to initiate movement on given slopes it is possible to estimate the thaw consolidation ratio appropriate to the field studies quoted in table 13. These are given in table 14 above.

The thaw consolidation theory has been applied to the prediction of thaw slope stability by McRoberts (1972), quoted by McRoberts and Morgenstern (1974) and McRoberts (1978). The slope model is illustrated in figure 53. The thaw front is considered to have penetrated to a depth x, so that the effective stress on the plane AA' after the dissipation of excess pore pressures is $\gamma'x\cos\beta$. The excess pore pressure u is given by:

$$u = \gamma'x\cos\beta \; (\frac{1}{1 + \frac{1}{2R^2}}) \qquad (4,22)$$

and therefore, pore pressure during thaw

$$u = \gamma_w x\cos\beta + \gamma'x\cos\beta \; (\frac{1}{1 + \frac{1}{2R^2}}) \qquad (4,23)$$

The effective stress is then,

$$\sigma' = \gamma x\cos\beta - \gamma_w x\cos\beta - \gamma'x\cos\beta \; (\frac{1}{1 + \frac{1}{2R^2}}) \qquad (4,24)$$

applying a static balance of forces, the factor of safety becomes

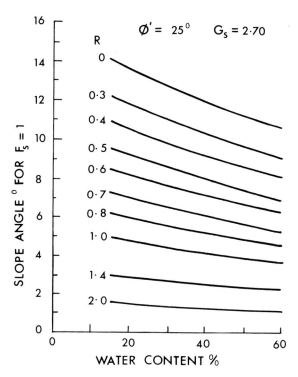

Figure 54 Solution of infinite slope analysis of thawing
slope in terms of the thaw consolidation ratio
R (McRoberts and Morgenstern 1974).

$$F = \frac{\gamma'}{\gamma} \left(1 - \frac{1}{1 + \frac{1}{2R^2}}\right) \frac{\tan\phi'}{\tan\beta}$$

assuming c' = 0.

This equation can be solved in terms of R and water
content, and McRoberts and Morgenstern provide a
graphical solution for a soil with $\phi' = 25^\circ$, and specific
gravity of particles 2.70 (figure 54). It is apparent
from figure 54 that for R values greater than 1 failure
can take place on very gentle slopes.

Hutchinson (1974) also recognises the importance of
enhanced water contents in frozen fine grained soils, and
the possibility that this water may be released more
rapidly than it can be expelled if thawing is
sufficiently rapid. Hutchinson states; 'where the rate
of thaw is sufficiently rapid relative to the rate of
consolidation of the thawed soil and where the shear
strength of this is sufficiently low, it is now suggested
that the active layer, wholly or in part, will tend to
slide downslope in a translational manner under conditions
that approximate to undrained' (Hutchinson 1974, p. 439).
Such displacements are considered most likely in clayey

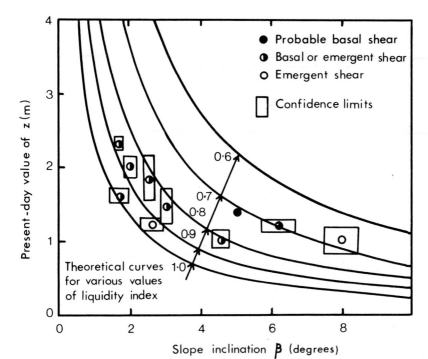

Figure 55 Theoretical relationship for undrained
conditions between thickness of sliding mantle
and slope inclination compared with field data
for shallow slab slides on London Clay
(Hutchinson 1974).

soils, and refer to slab sliding over well defined slip
surfaces. Although such failures have been described in
southern England, Hutchinson concedes that they are rare
in the modern periglacial zone.

Under drained conditions it is assumed that strength
is derived from cohesion since pore pressures reduce
$\sigma_n'\tan\phi'$ to zero. Adopting the infinite slope model
(figure 51), for conditions of limiting equilibrium,

$$z = \frac{2c_u}{\gamma} \, \text{cosec}\, 2\beta \qquad (4,25)$$

where z = vertical depth below ground level of shear
            surface
      c_u= undrained shear strength
      γ = bulk density of soil
      β = slope inclination

Using this relationship for fossil periglacial slope
material from London Clay, Hutchinson plots the
theoretical relationship between z and β for differing
values of liquidity index, and compares results with field
data (figure 55). This analysis indicates that as the
slope angle decreases the thickness of the unstable mass
tends to increase.

103

Concluding remarks

As was pointed out by McRoberts and Morgenstern (1974), from a geotechnical point of view the thaw consolidation model is entirely adequate in predicting the onset of unstable conditions in thawing slopes. However, quantitative modelling of the amount and nature of the resulting mass movement demands much greater knowledge of the properties and behaviour of 'frictional-viscous' soils. We also require more information on the geotechnical properties of sediments as they influence the mechanism of mass movement. In particular, more data on the significance of texture, the strength parameters ϕ' and c', and the index properties of soils to the mechanisms of mass movement might enable predication of when soil flowage and when slip failure are likely to occur. Indeed, as Carson (1978) has pointed out, more field work is necessary to establish whether viscous flow is the dominant mechanism of mass movement in the periglacial zone, or whether slip failures may play a more significant role than is currently recognised.

5. SOLIFLUCTION

Solifluction is the most widespread process of mass wasting above and beyond the tree line in both arctic and alpine environments. It may be associated with seasonal frost or permafrost. The importance of solifluction as a periglacial slope process is well illustrated by Holdgate et al. (1967) in their description of soils on Signy Island, South Orkney Islands. The authors describe solifluction as occurring on Signy Island 'on such a vast scale that it is often difficult to realize what is happening. Whole slopes are in motion, and it is not until an obstacle is encountered which causes buckling of the surface material that the extent of the motion is apparent' (p.60).

As noted in chapter 1, solifluction is the result of frost creep and gelifluction. The contribution of frost creep to the overall soil displacement is discussed in chapter 3.

Some physical properties of soliflucted soils

Texture

The soils of solifluction slopes in arctic and alpine areas consist mainly of sand or silt (figure 56) and generally have clay contents of less than 20%. However higher clay contents are reported from some sites, including two at Schefferville, Canada (Williams 1966) with clay contents 26% and 33%, a 'clay' layer beneath a stone stream in Signy Island (Chambers 1966) with 22% clay, and some of Washburn's sites in the Mesters Vig District, Greenland, with clay contents up to 30%

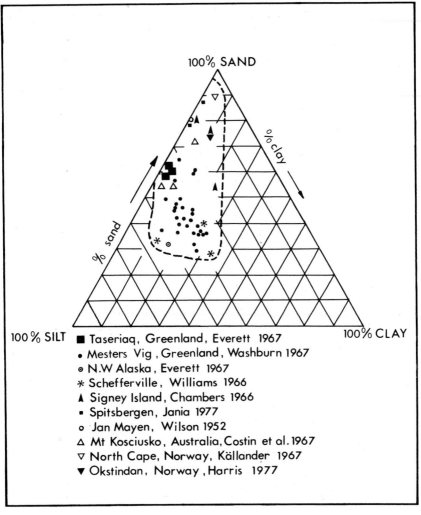

100% SAND

%clay

%sand

%

100% SILT

100% CLAY

■ Taseriaq, Greenland, Everett 1967
• Mesters Vig , Greenland, Washburn 1967
⊚ N.W Alaska , Everett 1967
✳ Schefferville , Williams 1966
▲ Signey Island , Chambers 1966
▪ Spitsbergen, Jania 1977
○ Jan Mayen, Wilson 1952
△ Mt Kosciusko, Australia, Costin et al. 1967
▽ North Cape, Norway, Källander 1967
▼ Okstindan, Norway , Harris 1977

Figure 56   Textural properties of soliflucted sediments.

(Washburn 1967). The infrequent occurrence of heavy
clays in the periglacial zone is due in part to the
preponderance of hard rocks, and the predominance of
mechanical over chemical weathering.   Glacial deposits
of various kinds often mantle bedrock and these are
largely silty or sandy in texture.   K. Graf (1973) has
analysed the grain size distributions of solifluction
material from Polar (figure 57a) and Alpine (figure 57b)
environments.   Again soils are shown to be silty or
sandy in character, with low clay contents.   The textural
envelope for all samples, including data from the Andes
and Kilimandjaro suggest that soils coarser than
Casegrande's frost susceptibility limits (non frost-
heaving soils) may also suffer saturated mass movements.

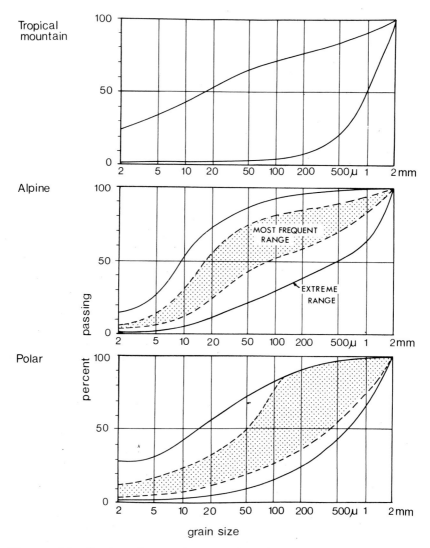

Figure 57   Textural envelopes of soils subject to
solifluction (Graf 1973).

The significance of textural variations to rates of
solifluction was investigated by Benedict (1970) in the
Colorado Front Range. At least five soil samples were
taken from seven instrumented sites and their silt/clay
contents plotted against movement rates at each site,
(figure 58). No correlation is apparent between texture
and movement rates, any relationship being obscured by the
effects of other, more important environmental factors.

Washburn (1967) observed that silt content may be a
critical variable in the susceptibility of a soil to
solifluction, plastic clays being too impermeable to

106

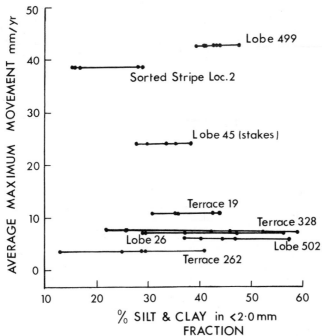

Figure 58   Relationship between texture and rates of mass
movement, Niwot Ridge, Colorado Front Range
(Benedict 1970).

support ice lensing, and coarse sand and gravel being too
well drained for saturated flow to occur, and non-frost
susceptible. Comparison of silt contents within his
monitored sites did suggest higher rates of movement
where silt contents were highest, but Washburn concluded
that of over-riding importance were soil moisture
conditions during thaw.

A factor which in some sites may encourage viscous
flow is the presence of mica flakes within the soil.
Holdgate et al. (1967) suggest that the 'abundant presence
of mica flakes in the Signy Island soils renders them
highly susceptible to solifluction, because movement tends
to align the flakes with their long axes parallel to the
direction of movement. Once this has taken place, lateral
movement between the flakes is facilitated and vertical
penetration of water is hindered' (p.60). Rapp (1960)
and Lundqvist (1962) in Sweden, and Källander (1967) and
Harris (1977) in Norway comment on the micaceous nature of
their soliflucted soils. Lundqvist noted that the
significance of mica-rich soils in promoting solifluction
was observed by Svenonius (1904).

Index properties and shear strength parameters

The sandy and silty soils which predominate in
solifluction sites described in the literature generally
have low liquid limits, low plasticity index, and largely

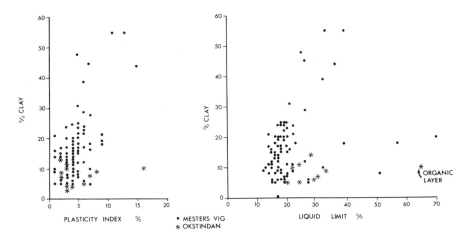

Figure 59   Relationships between clay content and Index
properties, Masters Vig, Greenland (Washburn
1967) and Okstindan, Norway (Harris 1972b).

frictional strength (table 15). The relative lack of
cohesion appears to be important in generating flow
displacements as opposed to slip failures in response to
high pore pressures. Lack of cohesion is associated with
low liquid limits in inorganic soils.

Figure 59 shows clay contents plotted against liquid
limit and plasticity index from Washburn's study of
solifluction at Mesters Vig, and Harris's study at
Okstindan, Norway. In both study areas liquid limits were
mainly around or below 25%, and both authors point out
that field moisture contents during thaw were often higher
than this. Jahn (1961) also observed moisture contents
above the liquid limit in Spitsbergen, but reports that
no soil flow took place until freezing in the autumn and
thawing during spring led to vigorous vertical movements
which disturbed the equilibrium of the soil.

Shear strength parameters are not widely reported by
geomorphologists studying solifluction in the field.
Williams (1966) estimated a value of ϕ greater than 30^0
for silty soils at Schefferville, Labrador-Ungava,
McRoberts and Morgenstern quoted values of $\phi_r' = 23^0$, $c_r' =$
0 for illitic clays in the Mackenzie Valley, and Harris
(1972, 1977) reports values of $\phi' = 29^0-33^0$, and $c_r' = 0$
for sandy solifluction soils in Okstindan, Norway (table
12).

Unit weight

The unit weight, or bulk density of near surface fine
grained soils in periglacial areas is likely to be
affected by frost heave in winter and resettlement during
summer (chapter 3). Where ice segregation occurs the
soil is liable to have a lower unit weight immediately

108

Table 15 Unit weights, soliflucted soils.

Location	Soil Texture	Average Dry Unit Weight (kg/m³)	(kN/m³)	Source
Taseriaq, W.	sandy silt	1630	16.0	Everett (1967)
Greenland		1750	17.2	
		1470	14.4	
Mesters Vig	clayey sand-silt	1870	18.3	Washburn (1967)
N.E. Greenland	silty sand	1540	15.1	
	gravelly sandy silt	2000	19.6	
	gravelly sandy silt	2140	21.0	
Okstindan N. Norway	silty sand	1660	16.3	Harris (1972)
		1580	15.5	
	gravelly silty sand	1510	14.8	

following thaw in spring than at the end of the summer
when settlement and compaction has taken place. Williams
(1959) illustrated such a variation in density on a
solifluction slope in central Norway where the average
dry unit weight in spring 1956 was 1370 kg/m³ (13.4kN/m³)
compared with 1460 kg/m³ (14.3kN/m³) in late summer. In
1957 the equivalent values were 1300 kg/m³ (12.75kN/m³)
and 1520 kg/m² (14.9 kN/m³) respectively.

Examples of unit weight determinations from active
solifluction slopes are given in table 15.

Field measurements of velocity distribution with depth

The distribution of soil movement with depth has been
most successfully investigated by the insertion into the
mobile soil of columns of plastic cylinders (Rudberg 1958,
1962, 1964, Dutkiewicz 1967, Benedict 1970, Harris 1972a,
b) or lengths of flexible plastic tubing (Williams 1966,
Price 1972, Harris 1977). The cylinders and tubes are
assumed to move with the soil so that their deformation
provides information on the distribution of soil movement
with depth.

Rudberg (1958) buried plastic cylinders 2cm high and

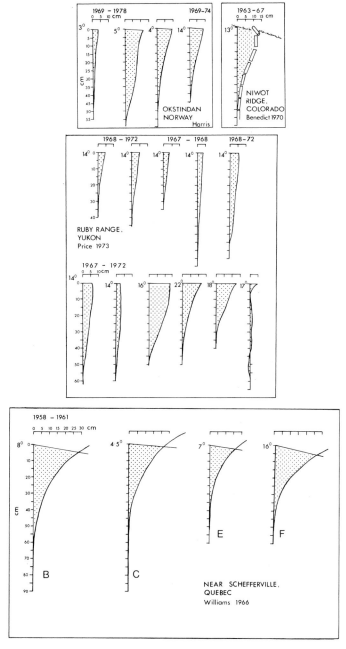

Figure 60   Velocity profiles due to solifluction.

2cm in diameter by lowering them through a pipe pressed
into the ground, which when removed left the cylinders in
a column within the soil. Harris (1972a,b, 1973)
describes the use of 5cm lengths of plastic tubing which
were threaded over a straight rod, before being pushed
into a vertical auger hole. Removal of the rod left a
column of tube sections firmly embedded in the soil.
Benedict (1970) used vertical columns of cement rods
installed by pushing them into an auger hole. Excavation
of these buried columns following a period of one, or
preferably more years reveals the distribution of movement
with depth. French (1974a) buried rectangular tinfoil
strips at various depths below the surface at Banks
Island in 1969, and these were re-excavated in 1972.
Movement of each tinfoil marker was measured against a
fixed base line.

Other methods of measuring subsurface soil movement
have attempted to overcome the basic problem associated
with the column method, that is, excavation necessarily
terminates the experiment. Williams (1962, 1966)
inserted lengths of flexible polythene tubing vertically
into the soil and measured their subsequent deformation
due to soil movement by means of a probe which fitted
inside the tubes. The probe consisted of a spring steel
strip on which were mounted a pair of strain gauges,
encased in epoxy resin. Bending of the probe led to a
change in electrical resistance of the strain gauges, so
that following calibration in the laboratory the
longitudinal shape of the polythene tubes could be found
by measuring the bending of the probe as it was pushed
down the tubes. Williams estimated errors to be less than
15%.

Examples of sub-surface velocity profiles due to
solifluction are given in figure 60, and corresponding
rates of soil movement are summarised in table 16.

From the diagrams it can be seen that movement
generally decreases with depth and dies out between 0.2
and 2.0 m below the surface. Profiles range from concave
downslope, to convex downslope, the two major factors
influencing their form being the distribution of
segregation ice in the frozen soil, and the thickness of
vegetation. The zone of maximum shearing associated with
thaw consolidation is likely to be where segregation ice
prior to thaw is thickest.

This has been emphasized by Rein & Burrous (1980) in
their large scale laboratory experiments to measure
subsurface displacements resulting from thaw consolidation
of frost susceptible soil at an inclination of 5°. Most
of the soil movement occurred in soil layers with ice
content in excess of 150% by weight. In one experiment
nearly 84% of the downslope surface displacement of 3.7cm
was concentrated in a single soil layer of unfrozen
thickness only 0.8 cm (3.2% of the total unfrozen soil

Table 16 Rates of solifluction measured by buried columns and flexible tubes.

Author	Location	Number of profiles referred to	Slope (degrees)	Maximum average annual surface (mm)	Maximum depth of movement (m)	Volume of soil passing monitored site per unit width of slope, per year (m³/m/yr)	
						max.	mean
Rudberg, S. (1964)	Norra Storfjäll Sweden	7	10-35	50	0.7	.01006	.0039
Benedict J.B. (1970)	Colorado Front Range	1	13	24	0.5		.0059
Harris, C. (1977) & unpublished	Okstindan Norway	3	4-14	16	0.65	.0033	.0026
Price, L.W. (1973)	Ruby Range, Yukon	10	14-22	35	0.66	.0071	.0032
Williams P.J. (1966)	Schefferville, Quebec	4	4-16	98	1.05	.0134	.0114

thickness). In its frozen state the ice content of this
layer was 250% by weight, and its thickness was 5cm.
Thaw consolidation led to a total surface settlement of
10cm, 4.2cm of which occurred due to consolidation of
this ice-rich layer.

Those profiles with strongly concave downslope form
are therefore probably associated with heaving ratios
which decrease from the surface downwards. Surface frost
creep may also contribute to relatively rapid surface
movements. Conversely, surface movement may be restricted
by a thick continuous mat of vegetation, to the extent
that rates of movement may increase for a short distance
below the surface, before decreasing to the base of the
moving mass (figure 60).

In Jan Mayen Island, Wilson (1952) observed that the
frontal banks of turf-banked lobes had a bulbous shape,
giving the impression of a mass of soil held back by a
skin of vegetation. Rudberg (1964) and Harris (1972) in
Scandinavia, and Price (1973) in the Ruby Range, Yukon,
also observed greater soil movement slightly below the
surface beneath thick vegetation cover.

In all cases movement of the soil is associated with
shearing distributed throughout the moving layer rather
than concentrated along discernable shear surfaces. From
the profiles of soil movement in figure 60 the volume of
soil passing the monitored point on the slope, per unit
width of slope, per year was found (table 16). Values
range from 0.0134m^3/m/year for a slope of around 8^0 at
Schefferville, to 0.00106 m^3/m/year on a slope of 10^0 in
Norra Storfjäll, Sweden. Maximum and mean values are
given in table 16.

Field measurements of surface rates of solifluction

Downslope displacement of the soil surface by
solifluction has been measured mainly by resurvey of
surface pebbles and boulders and of wooden pegs inserted
in the ground. Survey methods include theodolite
measurements from bedrock benchmarks (e.g. Washburn 1967),
steel tape measurements from bedrock benchmarks (e.g.
Rapp 1960), theodolite measurements from deeply inserted
stakes (e.g. Benedict 1970), steel tape measurements from
large boulders assumed to be stable (e.g. Price 1973) and
steel tape measurements from a taut wire or cord strung
between deeply inserted end-posts (e.g. Caine 1968).
Some indication of the rates of soil movement below the
surface may be obtained by inserting pegs to different
depths. The general conclusion that the rate of movement
decreases with depth is confirmed by pegs which tend to
tilt forward in a downslope direction in response to the
faster surface rates of movement.

Measured rates of surface movement are summarized
in table 17 where data are grouped into Alpine and Arctic

Table 17   Rates of surface movement, grouped into data
from (a) Alpine and (b) Arctic Regions.

(a)

Author	Location	Site Description	Slope (degrees)	Method of Measurement	Rate of Movement mm/yr Max	Min	Mean	Period of Study
Rapp A. (1960)	Kärkevagge, Sweden	vegetated solifluct-ion sheet	15-25	pegs	80	0	40	1953-59
Rudberg, S. (1964)	Norra Storfjäl, Sweden	vegetated solifluct-ion soil, below lobe	5	marked stones	65	0	20	1955-63
		vegetated solifluct-ion lobe	20		43	0	35	1955-62
		small poorly vegetated lobes	15		26	0	8	1957-63
		"	10		26	0	13	1957-63
		"	15		37	5	17	1957-63
Harris, C. (1972a)	Oskindan, Norway	turf-banked lobes	10-14	pegs	51	0	24	1970-71
			10-14		26	18	23	1970-71
Furrer, G. (1972)	Mt. Chavagl Swiss Alps	turf-banked lobes	3				29	
			9				69	
			10				42	
Pissart, A. (1977)	Chambeyron French Alps	small stone stripes	2	marked stones			10	1968-75
			16				50	1968-75
		fines in large sorted stripes	6-9		70	40		1963-75
Benedict J.B. (1970)	Colorado Front Range	stone-banked terrace	14	marked stones	7.3	1.5	4.2	1961-67
		stone-banked lobe	16-18		6.0	0.4	3.1	1965-67
		turf-banked lobe	11-12	pegs	10.9	1.9	2.0	1964-67

Table 17 (continued)

(a)

Author	Location	Site Description	Slope (degrees)	Method of Measurement	Rate of Movement mm/yr Max	Min	Mean	Period of Study
Benedict, J.B. (1970)	Colorado Front Range	turf-banked lobe	13		42.7	1.1	17.0	1962-67
		turf-banked terrace	6-7		22.8	0	9.8	1963-67
Caine, N. (1968)	Tasmania	solifluction terraces	8	marked stones			213.0	1974-75
			3				177.3	
			12				157.1	
			6				59.7	.
Price L.W. (1973)	Ruby Range, Yukon Territory Canada	southeast facing slope soli-fluction lobes	14	painted stones			16	1967-72
		east facing slope soli-fluction lobes	16	painted stones			14	1967-72
		north. facing slope largely un-differentiated	18	painted stones			9	1967-72

(b)

Author	Location	Site Description	Slope (degrees)	Method of Measurement	Rate of Movement mm/yr Max	Min	Mean	Period of Study
Barnett D.M. (1966)	Baffin Island	poorly veg-etated soli-fluction lobe fronts	5	pegs	52	7.5	13.5	1964-66
Jahn A. (1961)	Spits-bergen	solifluction lobes	5	pegs	120	5		1957-59
			5		60	0		
Jahn A. (1976)	Spits-bergen	vegetated solifluction slope	11	tilting of wooden piles			20-40	1943-74

Table 17 (continued)

(b)

Author	Location	Site Description	Slope (degrees)	Method of Measure-ment	Rate of Movement mm/yr			Period of Study
					Max	Min	Mean	
Holdgate M.W. et al. (1967)	Signy Island	solifluct-ion sheet			150			
Washburn A.L. (1967)	Mesters Vig, N.E. Greenland	solifluct-ion sheet, poorly veg-etated, wet	2.5	pegs	14	6	10	1956-61
			10.5		60	18.5	34	
			11.5		57	19.5	37	
		as above, dry	12.5		15	6	9	
			12.5		42	9	29	
			25		29	3.5	11	
		solifluct-ion lobe, wet	3-3.5		37	3	11	1957-61
			12		124	0	76	1957-59
French H.M. (1974a)	Sachs Harbour, Banks Island Canada	solifluct-ion sheet	2-4	tin foil markers	19.6	13.3	17.6	1969-72

environments. Movement rates of between 0 and 213 mm per annum are reported from Alpine areas, and from 0 to 124 mm per annum from Arctic areas. A notable feature of table 17 is that soil movement data for Alpine areas are mainly derived from solifluction lobes and terraces, while in the Arctic zone measurements of solifluction come from mainly undifferentiated solifluction sheets. This possibly suggests that solifluction is more widespread in Arctic regions underlain by permafrost, as opposed to the more localized occurrence of solifluction lobes and terraces in the Alpine zone. It is widely reported that solifluction rates are highest in the axial zone of the tread in solifluction lobes, and decrease towards the sides and towards the lobe front (e.g. Jahn 1961, Washburn 1967, Benedict 1970, Price 1973). Figure 61 illustrates surface movement rates recorded by Benedict (1970) on a small turf-banked lobe in the Colorado Front Range.

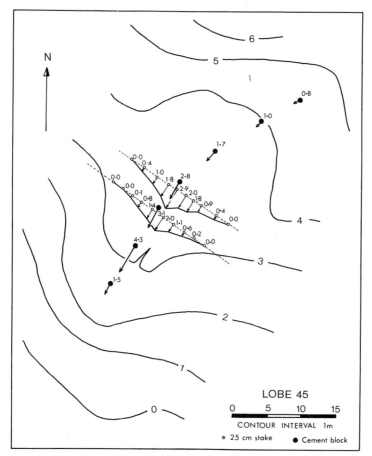

Figure 61 Rate of surface movement in a small turf-
banked lobe, Niwot Ridge, Colorado Front Range
(Benedict 1970).

Environmental factors affecting rates of solifluction

Soil thermal regime

Factors influencing the thermal regimes of periglacial
soils are discussed in Chapter 2. The rate of
penetration of the thaw plane in spring and summer, and
the period over which drainage is impeded by a frozen
subsoil are of particular significance to solifluction.
Considering first the rate of thawing, we have seen in
equation 4,20 that the thaw consolidation ratio R
increases with the rate of thawing, and in equation 4,21
that the excess pore pressure generated during thaw
increases as R increases. Hence rapid thawing is
associated with rapid release of frozen soil water and the
generation of the excess pore pressures necessary to
induce soil flow.

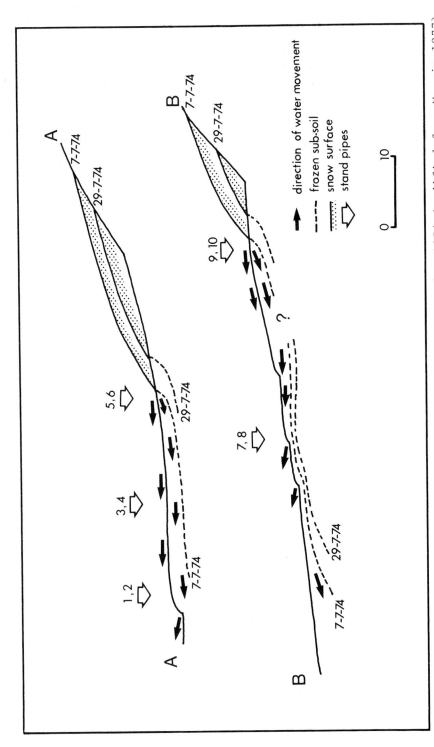

Figure 62 Solifluction slope in Okstindan, Norway during spring 1974 (modified from Harris 1977).

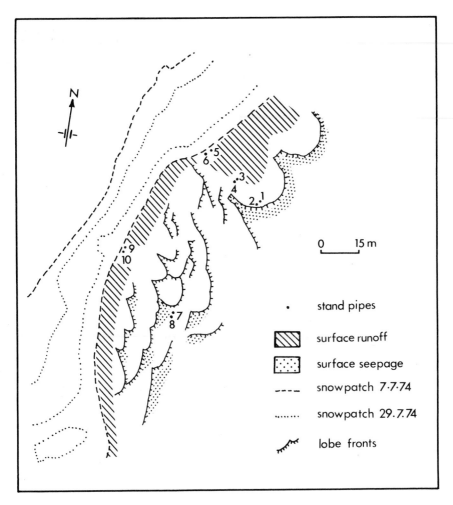

<u>Figure 63</u>   Surface runoff and surface seepage on a
                   solifluction slope in Okstindan, Norway,
                   during the spring thaw 1974 (modified from
                   Harris 1977).

    The importance of a frozen subsoil in impeding
drainage and thereby helping to generate soil saturation
is illustrated clearly by Harris (1972a, 1977) in his
study of solifluction lobes in Okstindan, Norway. During
the thaw period the frozen subsoil prevented vertical
drainage of meltwater released by snow and soil ice,
leading to lateral through flow over the frozen subsoil
surface, and a large volume of surface runoff (figures 62,
63). Where the frozen subsoil approached the surface,
seepage from the ground occurred. Water was observed
seeping rapidly out of the lobe fronts. Using soil
moisture tensiometers Harris showed that rapid soil
drainage occurred as soon as the frozen subsoil had

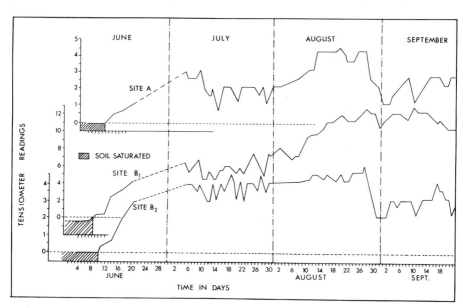

Figure 64   Tensiometer readings from a solifluction slope
in Okstindan, Norway during spring and summer
1970 (modified from Harris 1972a).

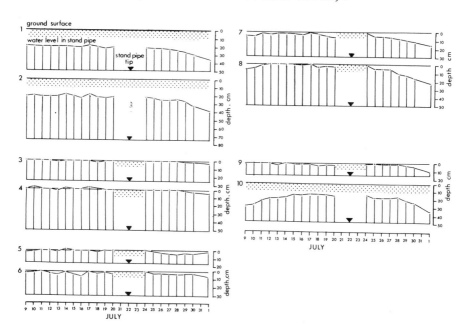

Figure 65   Water levels in stand pipes during thaw, 1974,
Okstindan, Norway. For location of stand pipes
see figure 62.

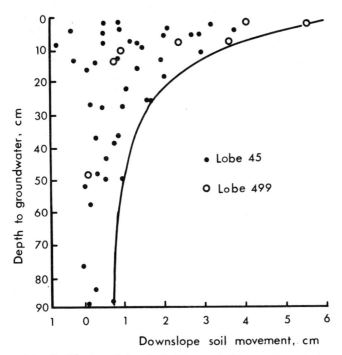

Figure 66   Relationship between surface movement and
             depth to water table during the autumn, Niwot
             Ridge, Colorado Front Range (Benedict 1970).

cleared in this non-permafrost area (figure 64).   The
persistence of a frozen subsoil will depend upon the
ratio between depth of winter frost penetration and rate
of spring thaw.   In permafrost areas drainage is
permanently impeded by the permafrost table.

Soil moisture conditions

     Soil moisture content is generally considered to be
the major factor influencing the occurrence of
solifluction.   As the slope stability analyses outlined
in chapter 4 show, high pore water pressures are required
to cause soil movement on the relatively gentle slopes
commonly associated with solifluction (see slope angles
quoted in table 17).   Sigafoos and Hopkins (1952) point
out that slow viscous flow is 'most conspicuous during
spring and early summer, when meltwater from snow and
ground ice wets the soil until it becomes a suspension'
(p.180).   Washburn (1967) and Harris (1972a) stress that
moisture content of the thawing soil outweighs the
influence of all other factors, and both authors point
out that in the silty and sandy soils commonly subject to
solifluction, moisture contents during thaw often exceed
the liquid limits.   In his study of solifluction lobes in
Okstindan, Norway, Harris demonstrated high moisture
contents during the thaw in 1974 using simple open stand-
pipes inserted in pairs down the monitored slope (figure
65).   On slope profile AA, a large gently sloping

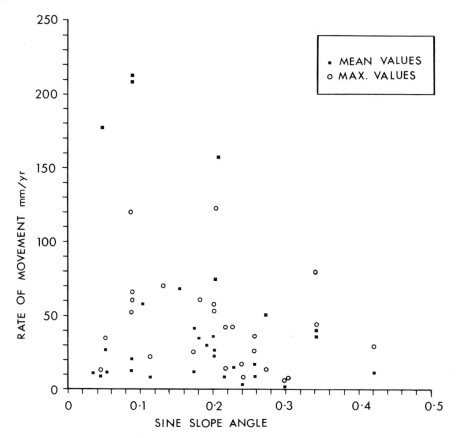

Figure 67 Relationship between surface movement rates and gradient, data from table 17.

turf-banked lobe, water levels in the stand-pipes, reflecting pore pressures at their tips, were at or near the surface through July, and artesian conditions prevailed for short periods in the mid-slope location. Near the lobe front the water table was lowered by seepage through the frontal bank. The mid-slope location in profiles BB showed similar high pore pressures, suggesting a water table near to the surface. Harris (1977) stressed the importance of snow melt in maintaining these high moisture contents.

Summer rainfall was observed to cause downslope soil movement in north-east Greenland by Washburn (1967), although it is not clear if this was due simply to increased soil moisture or to accelerated melting of the frozen sub-soil as relatively warm rainwater percolated into the active layer. Washburn's study also shows that soil movement by gelifluction is more susceptible to variation in moisture content than is frost creep. Consequently in 'wet' sites total solifluction movements

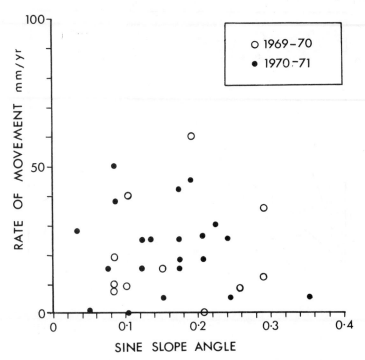

Figure 68   Surface movement rates against gradient for
             turf-banked lobes in Okstindan, Norway (Harris
             1972b).

were greater than in 'dry' sites, and the relative
contribution of frost creep to the total movements was
less.   In dry sites frost creep was the main cause of
downslope soil displacements (see chapter 3).

    Soil moisture conditions during autumn are also
important since the thickness of ice lenses developed as
the active layer freezes will depend largely on moisture
supply, providing the soil is texturally frost
susceptible.   Sigafoos and Hopkins (1951), describing mass
movements in Alaska, state that conditions are most
suitable for viscous flow in areas where frost heaving is
most intense during the autumn, and Benedict (1970) shows
that in the Colorado Front Range maximum soil movement
rates were recorded where the water table during the
autumn freeze was close to the surface (figure 66).   In
this study movement rates greater than 20 mm/yr occurred
only when the autumn water table lay within 20 cm of the
ground surface.

Gradient

    Since the downslope shearing force due to the soil's
own weight is proportional to the sine of the gradient,
one would anticipate a direct relationship between rates

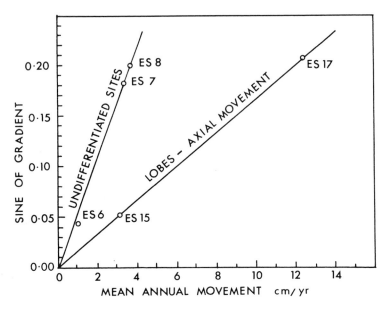

Figure 69   Relationship between surface movement rates and
             gradient, wet-sites, Mesters Vig, Greenland
             (Washburn 1967).

of movement and surface slope. However, plotting the
movement data in table 17 against the sine of the slope
gradients (figure 67) shows no apparent relationship
between movement rates and gradient. Clearly other
factors, noteably soil moisture conditions during spring
and summer, textural properties, and vegetation combine to
conceal any influence on solifluction rates that gradient
may have. Even when soil movement data collected from a
single small study area are plotted against gradient there
is no clear relationship (figure 68). However, when other
factors are fairly constant the influence of gradient
becomes apparent, as shown for wet sites in the Mesters
Vig District of north eastern Greenland by Washburn (1967).
Here the relationship between soil surface movement rates
and gradient appears to be linear, but higher movement
rates for a given gradient were observed at solifluction
lobe axes compared with undifferentiated solifluction
sheets (figure 69). This was largely due to the
correspondance between lobe axis location and lines of
maximum seepage through the soil downslope from snow,
patches, giving locally enhanced moisture contents.

    The range of gradients over which solifluction has
been reported is moderate. French (1974) describes
solifluction sheets on Banks Island in the Canadian Arctic,
moving down slopes of $2°$, and Washburn (1979) quotes
several examples of solifluction on slopes as low as $1°$.
The upper limit of slope gradient on which solifluction
can occur would appear to depend on the strength of the
soil and moisture conditions during the spring and summer

124

thaw. From conventional stability analysis, Harris (1977) concluded that in Okstindan, Norway, with the water table at the surface and lateral seepage, slope failure by landsliding rather than solifluction would occur on slopes in excess of $15°$ to $16°$, and in Spitsbergen, Chandler (1972) showed a threshold slope for landsliding of $18°$ to $20°$ under saturated conditions. In the Taseriaq area of West Greenland Everett (1967) describes rapid mass movements by slumping and mudflow on steeper slopes than those subject to solifluction. Solifluction sheets are reported on slopes mainly between $5°$ and $7°$, solifluction lobes on slopes between $5°$ and $15°$, and slumps and mudflows on slopes between $15°$ and $25°$. Large solifluction terraces in Alaska occur on slopes of between $5°$ and $15°$ (Sigafoos and Hopkins 1951), solifluction lobes mainly on slopes of between $20°$ and $25°$, and tundra mudflows occur in groups on slopes of $20°$ to $30°$. In northern Sweden Rapp (1960) reports solifluction lobes on slopes of between $15°$ and $25°$, and landslides and mudflows on slopes steeper than $25°$.

Perhaps the best illustration of the transition from solifluction on gentler slopes to rapid mudflows on steeper slopes in provided by Price (1969) in the Ruby Range, Yukon Territory. A valley with two distinct segments is described, the upper alb with gradient $14°$ to $17°$ and the lower glacial trough with gradient $25°$ to $29°$. There is an abrupt break of slope between these two segments. Solifluction lobes advance down the gently sloping alb, but at the break of slope they collapse, spilling their contents down over the steeper trough walls below. Collapsed lobes form distinct basin forms at the break of slope, from which mudflows extend down the steeper valley sides below.

It may be concluded therefore that where soil saturation occurs during thaw solifluction is the most important process of periglacial mass movement in unconsolidated sediments on slopes in the lower gradient range. At slopes in excess of $20°$ to $25°$ saturation of the soil leads to rapid mass movements of the landslide or mudflow type rather than solifluction. However, where soil saturation does not occur solifluction lobes may be observed on steeper slopes, even on slopes as steep as $38°$ (Harris 1976). Here soil creep rather than gelifluction is responsible for downslope movement and the development of lobes.

Topographic expression of solifluction

The topographic features produced by solifluction have been referred to above in the discussion of process. Three basic features have been recognised, solifluction sheets, solifluction terraces, and solifluction lobes. A simple classification by Benedict (1970) subdivides these features into sorted and unsorted as shown in table 18.

Table 18 Classification of solifluction features
(modified from Benedict 1970)

	No surface expression	Lobate	Terrace-like
Nonsorted	nonsorted sheet	turf-banked lobe	turf-banked terrace
Sorted	sorted sheet	stone-banked lobe	stone-banked terrace

Solifluction sheets are probably the least studied
solifluction feature, but possibly the most widespread,
particularly in the high Arctic where the absence of
vegetation enables solifluction to operate uniformly
(French 1976). In the sub-Arctic and alpine zones
vegetation cover and soil drainage are much more variable,
favouring the development of more localized lobes and
terraces. It should be noted that the term 'terrace'
includes both lobate and straight-fronted features, and
there appears to be no sharp dividing line between
terraces and lobes, the one often grading into the other.
In table 19 the main characteristics of solifluction
features, as reported by various authors, are summarized.
Here it can be seen that lobate forms are generally
smaller than terraces, with turf-banked lobes smaller
than the stone-banked variety.

Turf-banked lobes are reported as having risers up
to around 1 m in height, and treads of between 2 m and
50 m. They develop on slopes of between 4^0 and 25^0 or
more and several authors describe lobes on slopes below
late lying snow patches. Vegetation is generally thickest
on the fronts while the central areas of the lobe
surfaces are often less well vegetated. Stone-banked
lobes may have risers of up to 1 m or more, and in
Scotland Galloway (1961) describes risers of 5 m in
height. Treads of up to 60 m are reported. Stone-banked
lobes occur on slopes of similar gradient to those on
which turf-banked lobes develop. The major contrast
between the two is the relative abundance of coarser
material and lack of vegetation in the stone-banked lobes
compared with the turf-banked forms (figure 70). In
alpine areas stone-banked lobes and terraces tend to
occupy a higher altitudinal zone than their turf-banked
counterparts.

Turf-banked terraces range in height from less than
0.5 m to over 6 m, and have treads of a few tens of
metres, although in Alaska Sigafoos and Hopkins (1952)
report treads of up to 200 m.   Frontal banks may be
bulging and overhanging, or slope as gently as 30^0.

126

Table 19 Dimensions of solifluction features.

Feature	Author	Location	Slope (degrees)	Riser height (m)	Tread length* (m)	Tread width* (m)
Turf-banked lobes	Sigafoos & Hopkins (1952)	Alaska	20-25	0.3-1.5	6-50	3-10
	Williams (1957)	Central Norway	5-20	up to 1		
	Galloway (1961)	Highlands of Scotland		0.3-1.3	2-6	
	Lundqvist (1962)	Sweden	5-10	2-3		
	Rapp (1962)	N. Scandinavia	15-25	1	5-10	
	Dutkiewicz (1967)	Spitsbergen	2-6	0.05-0.1	4-6	
	Jahn (1967)	Spitsbergen	5-10	1.5	5	2-3
	Mottershead & White (1969)	Sutherland, Scotland	15-36	1(av.)	4.6(av.)	
	Benedict (1970)	Colorado Front Range	4-23	0.5-3.5	3-100	3-50
	Harris (1972b) (1976)	Okstindan, Norway	2-38	0.5-1.5	10-50	
	King (1972)	Cairngorms, Scotland	5-30			
	Price (1973)	Ruby Range, Yukon, Canada	14-18	1-6		
Stone-banked lobes	Sharp (1942)	St. Elais Range, Yukon	5-15	up to 0.6	up to 7.6	1.2-2.5
	Galloway (1961)	Highlands of Scotland		up to 4.5		
	Lewis & Lass (1965)	The Faroes	6+	0.6		

* Tread length measured downslope   Tread width measured along the contours

Table 19 (continued)

Feature	Author	Location	Slope (degrees)	Riser height (m)	Tread length (m)	Tread width (m)
Stone-banked lobes	Benedict (1970)	Colorado Front Range	12-24	up to 1		
	King (1972)	Cairngorms, Scotland	20-35		9-180	
	Harris (1976)	Okstindan, Norway	2-24	0.5-1.5		
	Jania	Spitsbergen	12(av.)			
Turf-banked terraces	Sigafoos & Hopkins (1952)	Alaska	5-15	1-6.5	30-200	100-1330
	Wilson (1952)	Jan Mayen	15-25	1-3	up to 10	
	Williams (1957)	Central Norway	5-20	up to 2	up to 50	
	Lewis & Lass (1965)	The Faroes	3-7	up to 0.46		
	Mottershead & White (1969)	Sutherland, Scotland	8-25	1.22(av.)	6.12(av.)	
	Benedict (1970)	Colorado Front Range	2-19 10(av.)	up to 4	up to 95	
Stone-banked terraces	Lewis & Lass (1965)	The Faroes	6+	1.5		
	Mottershead & White (1969)	Sutherland, Scotland	4-11	3.4(av.)	11.4(av.)	
	Benedict (1970)	Colorado Front Range	9-23 16(av.)	up to 2.7	up to 60	
Solifluction sheet	Smith (1956)	Spitsbergen	4-12			

Table 19 (continued)

Feature	Author	Location	Slope (degrees)	Riser height (m)	Tread length (m)	Tread width (m)
Soli-fluct-ion sheet	Everett (1967)	W. Greenland	5-7			
	French (1974)	Banks Island, N.W.T. Canada	3-20			

Turf-banked terraces are reported on slopes of between 2^0 and 20^0. Terrace-fronts often do not run parallel to the contours due possibly to structural controls of gradient, favourable micro habitats of south-facing terrace fronts, dominant wind direction leading to variation in vegetation thickness, and variations in snow distribution. Stone-banked terraces have frontal heights of similar dimensions to those of turf-banked terraces, and treads up to 60 m long.

The frontal banks of lobes and terraces result from variations in the rate of solifluction across a slope, material from above tending to accumulate in the zones of slower soil movement. The sorting of coarser debris to the frontal banks of stone-banked lobes and terraces may result from initial frost heaving of stones to the surface followed by their movement downslope at a faster rate than the underlying finer subsoil, since rates of solifluction generally increase towards the surface. The superficial coarser debris will therefore reach the frontal zone relatively quickly and there accumulate to form the stone bank. Such an explanation is supported by stone orientation measurements (e.g. Lundqvist 1949) which show elongated clasts with downslope orientation on the tread but orientations across the slope, parallel to the front, in the riser. Within the riser the b-(intermediate) axes of these oriented stones tend to dip against the slope in an imbricate fashion.

Excavation of turf-banked lobes in the Okstindan Mountains of Norway revealed marked concentrations of stones and boulders in the frontal banks beneath the vegetation cover (figure 73) (Harris 1972). Similar concentrations in the fronts of turf-banked terraces are reported by Williams (1957) in central Norway where 'frequently there are many stones and boulders arranged peripherally under the surface of the frontal bank' (p.43) and by Everett, in West Greenland. This suggests that there is no sharp distinction between turf-banked and stone-banked solifluction features but rather a

Figure 70   Turf-banked lobes, Okstindan, Norway.
Lobe fronts are between 0.5 m and 1 m in
height.

Figure 71   Stone-banked lobe, Okstindan, Norway.
Note rucksack for scale.

continuum between them, depending on the degree of
vegetation development.

Internal stratigraphy of terraces and lobes, and rates of
frontal advance

Excavation of solifluction lobes and terraces where
the slope is vegetated almost always reveals buried
organic-rich horizons extending back from the base of
the riser (Hanson 1950, Sigafoos and Hopkins 1952, Wilson

130

Figure 72   Trenched stone-banked lobe, Okstindan,
Norway. Note the concentration of stones and
boulders at the surface, with matrix-dominated
sediment beneath.

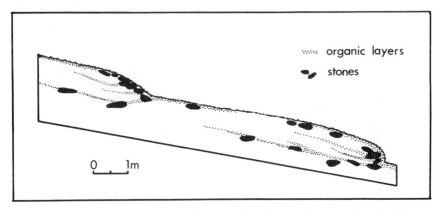

Figure 73   Section through turf-banked lobes, Okstindan,
Norway (Harris 1972b).

1952, Williams 1957, Lewis and Lass 1965, Benedict 1966,
1970, Everett 1967, Jahn 1967, Dylik 1969, Price 1970,
Costin 1972, White and Mottershead 1973, Worsley and
Harris 1974, Furrer and Bachmann 1972, Mottershead 1978,
Ellis 1978a,b, Alexander and Price 1980). An example of
lobe stratigraphy is shown in figure 73. The organic
layers are generally interpreted as pedological soil
organic horizons buried by the progressive advance of a
terrace or lobe over the ground surface infront of it.
Harris and Ellis (1980) observed a well developed micro-
podzol which passed beneath the front of an advancing
turf-banked terrace in Okstindan, Norway with no
apparent disturbance of its structure. However, buried

organic horizons are often contorted, and may comprise several distinct layers, particularly beneath turf-banked lobes and smaller turf-banked terraces. Whisps of organic material also commonly extend downwards and backwards from the modern soil A horizon.

Soil movement data reported earlier in this chapter indicate most rapid movement rates at the surface, with movement decreasing with depth and generally dying out at about 0.5 m below the surface. Since lobes and terraces are commonly more than a metre in thickness it is apparent that the whole lobe or terrace is not moving forward *en masse* , but rather, the superficial few tens of centimetres are advancing over an essentially stable subsoil. The complex nature of the buried organic layers suggests that the frontal advance may take place in several ways. Benedict (1970) describes a lobe where the A horizon of the modern soil can be traced down over the lobe front and then back beneath the front in an inverted position, as part of the buried organic layer. Below the inverted buried soil is a normal buried soil representing the ground surface overridden by the advancing lobe. Thus, this lobe advances by a 'rolling over' process, the surface soil behaving rather like a caterpillar track, being inverted and buried as the lobe moves forwards. Similar folding over of surface vegetation at the front is described by Troll (1944), Washburn (1947), Wilson (1952) and Ellis (1978). Everett (1967) and Price (1970) suggest that the disruption of the buried organic layers may be due to shearing within the saturated soil mass, while the measurement of stone dips in a lobe front in Okstindan, Norway led Ellis (1978) to infer that frontal advance took place spasmodically as a result of rupture of the retaining turf bank through which occurred rapid viscous flow of saturated sediment, depositing a thin sheet over the adjacent downslope area. Repeated temporally spaced breaches could lead to the succession of buried organic layers observed by Ellis beneath the lobe front. Finally, Wilson (1952) in Jan Mayen describes the downslope advance of vegetated terrace fronts due to the build up of pressure behind the 'skin' of vegetation which gradually 'gives' by stretching. Creeping vegetation such as *Salix herbacea* becomes stretched in a downslope direction by frontal advance, and there is no overriding of the vegetation and soil in the path of the advancing terrace.

Clearly organic material buried by the advance of solifluction lobes and terraces may be dated by radiocarbon assay. If it is assumed that no new carbon is added to the organic layers following their burial, and the age of the organic material prior to burial (when it formed the A horizon of a soil) is known, the date of burial at any point along an organic layer can be found. Where samples are dated at regular intervals upslope from the terrace or lobe front, the rate of frontal advance

during the period of terrace or lobe formation may be
deduced. However, dating of buried soils is subject to
many sources of error, and Matthews (1980) lists the
following major problems:
(i) the complexity of soil organic matter and the
difficulties of identification and separation of organic
fractions of differing age;
(ii) the increase in age with depth in many soil profiles,
a consequence of mobility and/or accumulation of organic
matter;
(iii) the non-comparability of different soil types with
respect of age;
(iv) an unknown degree of within-type variability in soil
age;
(v) the maturity of the soil at the time of its burial,
and whether or not it was in equilibrium;
(vi) contamination of modern analogue soils by 'bomb'
carbon, making it difficult to estimate their ages prior
to burial;
(vii) the susceptability of soil organic matter to
contamination from lithogenic carbon, 'bomb' carbon in
percolating water, and post burial rootlet penetration;
(viii) the possibility of disturbance and erosion of soils
during burial;
(ix) the possibility that biological and physico-chemical
transformation within the soil continue after burial.

These problems largely relate to establishing the
radiocarbon age of the soil organic horizon prior to its
burial by the advancing lobe or terrace front. This age
must be deducted from the radiocarbon date obtained from
the buried soil inorder to estimate its time of burial.
During soil formation organic matter is added to the
surface and over time decays until it is lost through
oxidation or leaching. The A horizon therefore has an
average 'age' referred to as the 'apparent mean residence
time' for organic material (Sharpenseel and Schiffman
1977). If the soil is mature and in equilibrium this
apparent mean residence time should be constant.

The few dates which are available of modern A horizons
in periglacial areas range from around 50 radiocarbon
years to around 500 radiocarbon years (Østrem 1965,
Benedict 1966, Ellis 1979, Alexander and Price 1980), but
bearing in mind the problems listed above, and the
additional problem of calibrating radiocarbon dates to
account for fluctuations in atmospheric ^{14}C (e.g. Suess
1970), any estimate of apparent mean residence time must
be treated with caution.

In a review of published radiocarbon dates from
organic layers buried by solifluction lobes and terraces
Benedict (1976) shows that sites in Greenland (Everett
1967), Yukon Territory (Price 1970), Alaska (Hamilton,
quoted by Benedict), Australia (Costin et al. 1967),
Scotland (White and Mottershead 1973) and Norway
(Worsley and Harris 1974) suggest long-term average rates

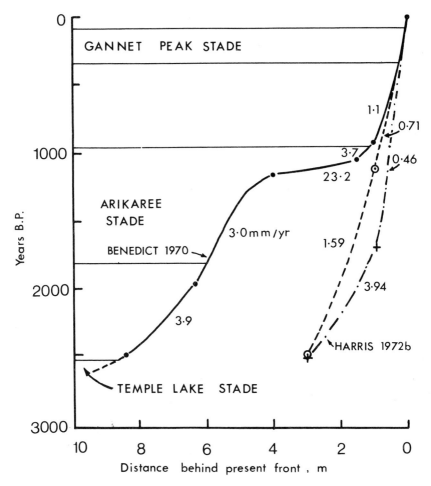

Figure 74   Rates of lobe front advance from radiocarbon
dating of buried organic material.

of frontal advance of between 0.6 and 3.5 mm per year.
However, from the examples illustrated in figure 74 it
can be seen that somewhat higher rates of frontal advance
may have occurred in the past. Reference to tables 16
and 17 shows that frontal advance of terraces and lobes
is much slower than solifluction rates recorded on their
tread surfaces. Measurements of surface rates of mass
movement in turf-banked lobes in the Ruby Range by Price
(1973) confirm greater rates of downslope movement on the
tread surface than at the fronts, and Benedict (1976)
suggests that estimates of sediment transport by lobes
and terraces will be an order of magnitude too high if
they are based purely on velocity measurements made in
midtread positions.

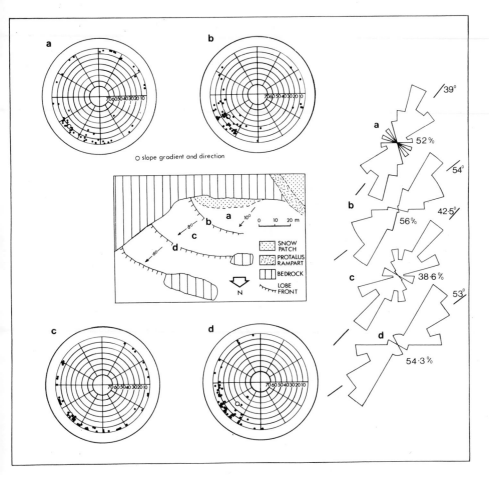

Figure 75   Macrofabrics from stone-banked lobes in
             Okstindan, Norway (Harris, unpublished).
             Fabric (d) in this figure is from the trenched
             lobe illustrated in figure 72. Computed
             preferred orientations and vector magnitudes
             are indicated.

Fabric analysis of soliflucted sediments

Fabric elements may be defined as any components of a
sediment which behave as single units with respect to an
applied force. Fabric elements usually considered in
fabric analysis are either pebbles (macrofabric) or sand
grains (microfabric). Shearing within the flowing soil
mass during solifluction leads to the alignment of
elongate fabric elements (pebbles and sand grains)
parallel to the direction of flow, which is generally
parallel to the slope azimuth.

Downslope orientation of pebble long axes as a result

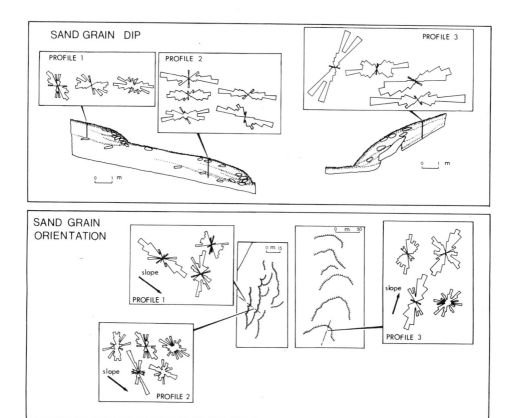

Figure 76   Microfabrics from turf-banked lobes and a turf-
banked terrace, Okstindan, Norway (data from
Harris and Ellis 1980).

of solifluction is widely reported (see Washburn 1979, p.
217, and Smith 1956, Ragg and Bibby 1966, Furrer 1972,
King 1972).   Elongate pebbles also tend to dip roughly
parallel to the surface, or at an angle slightly less
than the slope angle, in an imbricate fashion relative to
the slope (e.g. Smith 1956, Benedict 1970 fig. 51).
Figure 75 illustrates macrofabrics from stone-banked lobes
in the Okstindan Mountains, Norway.   The lobes were
developed in till of the last major Scandinavian ice sheet,
consisting of sub-angular to sub-rounded boulders and
pebbles set in a silty sand matrix.

Microfabric analysis of solifluction sediments is less
widely reported, the technique involving removal of

undisturbed oriented samples from the field, laboratory
impregnation with a suitable low-viscosity resin, and
making oriented thin sections from the impregnated samples
prior to microscopic examination (Benedict 1969, Harris
and Ellis 1980). Vertical thin sections described by
Benedict (1970, fig. 21 and fig. 51) showed that within
both a turf-banked lobe and a stone-banked lobe in the
Colorado Front Range sand grains tended to lie parallel
to the surface, or in an imbricate fashion relative to the
surface except at the lobe fronts where grains were not
strongly orientated. In the stone-banked lobe strength
of sand grain orientations was similar to that of pebble
orientation in the vertical plane. Harris and Ellis
(1980) similarly show a strong preferred dip of sand
grains in vertical thin sections taken from turf-banked
lobes and a lobate terrace in Okstindan, Norway (figure
76), with grains tending to dip parallel to the surface.
Sand grain orientations in horizontal thin sections
described by Harris and Ellis were approximately parallel
to the slope azimuth (figure 76), but were less strongly
orientated than in the vertical plane. A comparison of
figures 75 and 76 also suggests that the orientation of
sand grains parallel to the slope azimuth as observed in
the horizontal thin sections is less strongly developed
than the orientation of elongate pebbles. The
microfabrics therefore appear rather more variable, and
are generally more isotropic than the macrofabrics.

Micromorphology

    In addition to the measurements of sand grain
orientations mentioned above, Harris and Ellis (1980) also
described the organisation and distribution of finer
grained (mainly silt) components of the soil in turf-
banked lobes and terraces in the Okstindan Mountains,
Norway. This finer material is referred to as matrix,
defined as mineral material which is capable of being
translocated, and including grains up to 60μ in diameter.
Two profiles in turf-banked solifluction lobes and one in
a larger lobate terrace (figure 76) were sampled.

    In all three profiles matrix material was observed
forming coatings on sand grains. Below approximately 0.5
m the upper surfaces of grains were the most frequent
locations for matrix accumulations, which formed smooth-
surfaced streamlined cappings (figure 77). At shallower
depths matrix coatings also occurred on the lower and side
surfaces of sand grains. Maximum thicknesses of up to
2 mm were observed on the larger skeletal grains. Harris
and Ellis suggest that at this site the rapid vertical
drainage of the solifluction lobes and terraces following
clearance of seasonal frozen ground in summer leads to
silt being washed down the profile. This silt is
intercepted by the upper surfaces of sand grains where it
forms coatings. In the upper 0.5 m of soil, solifluction
may disturb the orientations of sand grains, so that
coatings may not occur only on the upper surface, but on

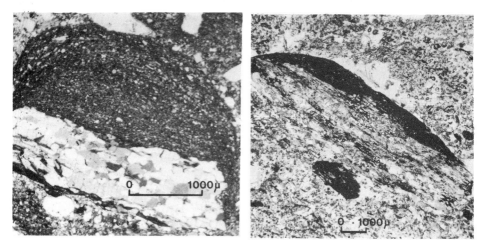

Figure 77   Silt cappings on upper surfaces of sand grains,
turf-banked lobes, Okstindan, Norway. (a) 25 cm
depth, (b) 10 cm depth.

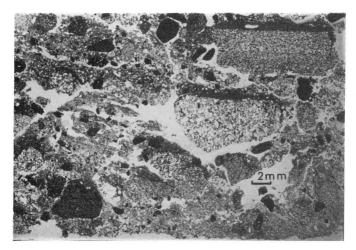

Figure 78   Silt cappings on upper surfaces of gravel-sized
fragments, solifluction deposits, Taf Fechan,
Brecon Beacons, South Wales.

other surfaces as well. Harris (1980) in a preliminary
report on fossil solifluction deposits in South Wales
shows similar silt cappings on sand and gravel sized
material (figure 78) at sites in the South Wales Coalfield
and the Brecon Beacons. He suggests that such
micromorphological elements might be useful in identifying
sediments which have suffered solifluction in the past.
Harris and Ellis (1980) also observed silt concentrated
into somewhat flattened patches of matrix enrichment.
These lay roughly horizontally, and became increasingly

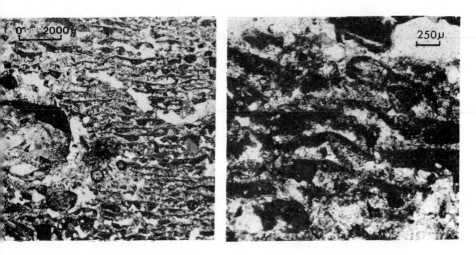

Figure 79   Lenticular silt concentrations, 50 cm depth, turf-banked lobe, Okstindan, Norway.

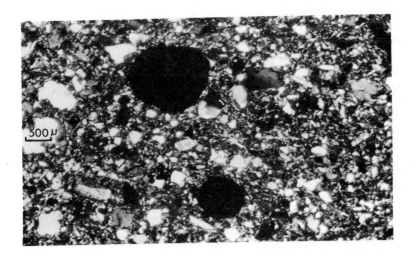

Figure 80   Vesicular pores, 25 cm depth, turf-banked lobe, Okstindan, Norway.

more flattened with depth, so that below about 0.8 m a large proportion of the matrix was concentrated into horizontal streaks or layers, separated by coarser detrital grains (figure 79). Such lenticular or banded silt concentrations have been described elsewhere in cold environment soils, and are not considered to be produced by solifluction, but rather by the development of segregation ice during freezing. The lenticular pores left by melting ice lenses are occupied by silty soil water, and form the locii of silt enrichment following the expulsion of excess water and the closure of the pores by soil resettlement. Active solifluction in the near surface layers may prevent the formation of well developed lenses of silt enrichment.

Finally, relatively large rounded vesicular pores were observed in the upper 0.25 m of the profiles investigated (figure 80). They were on average around 500μ in diameter, but pores up to 3 mm were noted. Again these features are not considered to result from solifluction, since they have been reported in periglacial soils not subject to solifluction. It is considered that these vesicles are formed during soil freezing as a result of expulsion of air dissolved in the soil water, the air being trapped as bubbles within the frozen soil mass (Fitzpatrick 1956). Harris and Ellis suggest that these are essentially transient features produced annually by the winter freeze and destroyed during soil resettlement and saturated flow during thaw.

Summary and concluding remarks

Solifluction is a convenient term to describe the downslope saturated flow of unconsolidated sediments in periglacial areas. It includes two processes, gelifluction and frost creep, processes which are often impossible to separate in field studies. The relative importance of gelifluction and frost creep varies from site to site and from year to year, but it appears that gelifluction is more sensitive to soil moisture conditions during thaw, so that frost creep tends to be more important in drier sites. Solifluction is considered to result from thaw consolidation of non-cohesive soils under saturated conditions. Soil saturation is promoted by the impedence of drainage by permafrost or seasonal frozen ground, and soil water is derived from snow melt and thawing ice lenses. Rates of solifluction show considerable within-site variation, soil moisture conditions during thaw being the most important controlling factor. Solifluction may take place on low angled slopes down to 1^0 or even less, but on slopes in excess of around 25^0 it tends to be replaced by more rapid mass movements such as mudflows and slides.

The most widespread topographic expressions of solifluction in Arctic environments are smooth straight or convexo-concave slopes mantled with solifluction sheets, but in Alpine environments greater variation in solifluction rates appear to occur, producing solifluction lobes and terraces on hillsides. These lobes and terraces tend to be vegetated at lower altitudes but unvegetated with larger stones and boulders concentrated in the frontal zones at higher altitudes. Solifluction leads to the orientation of elongate fabric elements in the soil parallel to the direction of flow, and produces a preferred dip parallel, or slightly imbricate to the ground surface.

Despite the apparent mass of data given in table 17
there remains a lack of long term studies of the rates of
solifluction. In addition, alpine areas have generally
received more attention from geomorphologists measuring
rates of mass movement than have arctic areas. Given the
inherent variability of rates of mass movement both
spatially and temporally, there is a danger of over-
reliance on a few well-reported studies. Furthermore,
this danger is increased by the tendency for field data
to relate specifically to slopes where mass-wasting is
clearly active, such as those exhibiting well-developed
solifluction lobes and terraces. Much more data is
required, particularly from the arctic where such clearly
visible topographic features may not be present, but
where mass-wasting is none-the-less an important agent of
sediment movement on slopes.

6. RAPID MASS MOVEMENTS

Skinflows and active layer glides

The term skinflow is used by McRoberts and Morgenstern
(1973) to describe rapid periglacial mass movements
involving the detachment of a thin veneer of soil and
vegetation and its subsequent downslope displacement over
a planar surface, generally the permafrost table. The
authors suggest that flow is the dominent process of movement.
An alternative and often used term for such mass movements
is mudflow, but topographically similar failures have
also been described where sliding over a distinct shallow
shear plane is proposed as the chief mechanism of
movement. These slip failures are termed 'active layer
glides' by Mackay and Matthews (1972). Infact it is often
difficult to quantify the relative contribution of slide
and flow in a given failure and there appears to be a
continuum between slip-dominated and flow-dominated mass
movements. Failures of the skinflow and active layer
glide type are generally shallow in relation to their
depth, and extend downslope as long narrow ribbon-like
failures (figure 81). Alternative nomenclature used to
describe these shallow active layer failures is summarized
in table 20.

Skinflows are relatively common in the arctic tundra
zone (French 1976) and also in the subarctic boreal forest
where the resulting disruption of tree cover makes such
failures clearly visible from the air (see for instance
figure 86, p. 366, McRoberts 1978). McRoberts and
Morgenstern describe skinflows developed in soils derived
from frost weathered shales in the Root River Valley,
N.W.T. Canada. They occur on north-facing slopes of
inclination 25°, are some 305 m in length, and up to
0.9 m deep. Similar skinflows described by the same
authors occur in colluvium on the Hanna River N.W.T., with
surface gradients of only 6°.

Figure 81   Skinflows in Christopher Shales, South Banks
Island, Canadian Arctic.   Photograph courtesy
of H.M. French.

In the Mackenzie River Valley, N.W.T., Canada, Mackay
and Matthews (1972) describe two types of shallow active
layer failure;   active layer glides, and mudflows (table
20).   Active layer glides involve the detachment of a
segment of the active layer, which slides downslope over
the permafrost surface, suffering little internal
disturbance.   French (1976) illustrates a similar
failure in Banks Island, in the Canadian Arctic (figure
89, p.147, French 1976) where the fine grained sediments
which make up the shallow active layer are particularly
susceptible to active layer failures of this type.   French
suggests that failures occur mainly on slopes of 15° or
more.

Mudflows are generally considered to be flow-
dominated failures (table 20) and involve sediments with
extremely high water contents.   Mackay and Matthews (1972)
stress that mudflows are associated with the melting of
ice-rich sediments, so that the scar left by the flow may
have a greater volume than the mudflow deposit, the
volume difference reflecting the amount of excess ice in
the soil prior to thaw.

In the Kärkevagge area of northern Scandinavia, Rapp
(1960) observed mudflows in silty sand till which occur
during the spring thaw, when snow melt and melting ground
ice enhances the water content of the partially thawed
active layer, and later in the season, following heavy
rainfall.   The flows extend downslope as ribbons of debris
having rounded lobate fronts and flanked by ridges or
levées 3-4 m wide and up to 1 m high.   The uppermost parts
of the levées are bouldery, fines having been washed out,
and elongate rock fragments tend to be oriented downslope.
In the mudflow lobes, however, stone orientations are

Table 20   Rapid mass movement of the skinflow/active
            layer glide type.

Proposed	Mechanism of Movement	
Mainly flow	Intermediate	Mainly slide
Skinflow[1]	Debris flow[4]	Active layer glide[6]
Mudflow[2]	Detachment failure[5]	
Earthflow[3]		

[1]McRoberts and Morgenstern (1973)
[2]Sharp (1942), Sigafoos and Hopkins (1952), Rapp (1960), Mackay and
 Matthews (1972)
[3]Holmes and Lewis (1965)
[4]Anderson et al (1969), Jahn (1976)
[5]Hughes (1972)
[6]Mackay and Matthews (1972)

generally transverse to movement. The troughs extending
downslope between the levées often provide channels for
subsequent runoff, and as a result are enlarged to form
gullies 1 m to 2 m deep and 2 m to 6 m wide. The average
gradient for mudflows in the Kårkevagge area is 30°,
suggesting that excessively high pore pressures are not
necessary to initiate failure. Rapp's description of
mudflows corresponds closely to those of Sharp (1942) in
the St. Elias Range, Yukon, Holmes and Lewis (1965) in
the Mt. Chamberlin area of Alaska, Anderson et al. (1969)
in Northern Alaska, and Jahn (1976) in Spitsbergen
although in these subarctic and arctic permafrost areas
failures on much gentler slopes are common.

Nature of sediments

    Skinflows and active layer glides are reported in
sediments ranging from bentonite clays in Northern Alaska
(Anderson et al. 1969), to coarse grained till (McRoberts
and Morgenstern 1974). Rapp (1960) observed mudflows in
a silty sand till comprising 38% gravel, 48% sand and
14% silt/clay, and Holmes and Lewis (1965) in Alaska
describe similar flows in till with 61% gravel, 28% sand
and 11% silt/clay.

Mechanism of movement

    The mode of displacement of sediment in skinflows is
considered by McRoberts and Morgenstern (1974) to be akin
to solifluction, the major difference being the rate of
movement. Solifluction is the characteristic type of mass
wasting on many slopes in the periglacial zone, occurring
annually under average conditions (McRoberts 1978).

143

Skinflows, however, are sudden, sporadic failures of short duration, which often result from an abnormal event, such as fire, heavy rain or high temperatures, leading to unusually rapid and deep thawing of the active layer. French (1976) suggests that when the spring thaw is late and therefore rapid, skinflows are likely to occur. Sigafoos and Hopkins (1952) stress the essential similarity between slow and rapid viscous flow (solifluction and mudflow respectively). They suggest that severe frost heaving is important in initiating skinflows of the mudflow type, because it not only leads to high soil water contents during thaw, and consequent high pore pressures during consolidation, but also disrupts the binding turf mat and facilitates the detachment and rapid displacement of units of the active layer.

Anderson et al. (1969) observed that the floor and walls of debris flow channels in northern Alaskan clays are smoothed and fluted indicating sliding as well as flow, producing a silkensided slip surface. The cohesive strength of the clay is probably responsible for this type of failure. Active layer glides described by Mackay and Matthews (1972) in the Mackenzie River Valley, Canada, are also associated with sliding over a distinct slip surface which is likely to lie close to the boundary between the active layer and the permafrost. This boundary is commonly ice rich and therefore releases large amounts of excess pore water on thawing. Again some kind of trigger initiating unusually rapid thawing of the permafrost is considered necessary to initiate sliding, and Mackay and Matthews suggest heavy summer rainfall and destruction of the vegetation cover by fire as possible mechanisms. The lack of disruption of the sliding segment of active layer is due to the binding effect of plant roots.

In all cases skinflows and active layer glides are described as short duration rapid failures. Holmes and Lewis (1965) observed a skinflow in Alaska for instance, which was initiated by rapid thawing in a very mild spell during a milder than normal summer, where movement lasted for three days.

Shallow active layer failures are reported on slopes ranging from a few degrees to 30° or more. On steeper slopes they may merge to form broad sheets of instability (McRoberts and Morgenstern 1974, Sigafoos and Hopkins 1952). On steeper slopes, particularly when the binding effect of vegetation is reduced, excessive pore pressures may not be necessary to initiate failure, but on gentler slopes high pore pressures generated by rapid thawing of ice rich soils and slow dissipation of the excess pore water must play a major role in initiating these failures.

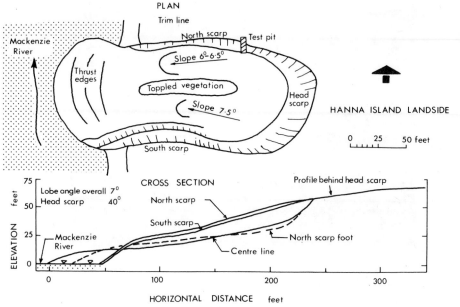

Figure 82   Bimodal flow, North West Territories, Canada
             (McRoberts and Morgenstern 1974).

Bimodal flows

Bimodal flows (McRoberts 1973) are slope failures
which develop when ice-rich permafrost thaws. They occur
on river valley sides, coastal slopes, and around
thermokarst depressions. They occur only in the presence
of permafrost. The term bimodal flow indicates two
distinct sectors (McRoberts and Morgenstern 1974,
McRoberts 1978), an upper, steeply inclined headscarp,
and a lower less steep tongue or lobe extending out and
away from the scarp (figures 82 and 83). These zones
also have distinctive modes of mass wasting, with ablation
of the permafrost on the scarp releasing debris which
falls, slides, flows or is washed down to the lobe, where
it flows away. Bimodal flows have been described in the
Canadian Arctic Islands (Victoria Island, Washburn 1947;
Ellef Ringes Island and Axel Heiberg Island, Lamothe and
St-Onge 1961; Garry Island, Kerfoot 1969; Banks Island,
French and Egginton 1973, French 1974b); the Mackenzie
Valley (Hughes 1972, McRoberts and Morgenstern 1974), the
Mackenzie Delta (Mackay 1966), and East Siberia (Czudek
and Demek 1973). They have been referred to as slumps
(Mackay 1966), ground ice slumps (French 1976), thaw
slumps (Washburn 1979), mud slumps and mudflows (Kerfoot
1969), retrogressive thaw flow slides (Hughes 1972) and
thermo-erosion cirques (Czudek and Demek 1970).

Lamothe and St-Onge (1961) provide a vivid
description of the cycle of development and decay of a
bimodal flow in the Isachsen area, Ellef Ringes Island.

145

Figure 83   (a) Bimodal flows caused by lateral stream
            migration undercutting valley sides, East Banks
            Island, Canadian Arctic.
            (b) Scarp of bimodal flow, East Banks Island.
            Photographs courtesy of H.M. French.

Here a silty active layer is underlain by massive ground
ice up to 0.55 m in thickness. This ground ice becomes
exposed initially as a result of the slumping and gravity
fall of material where a river rapidly undercuts and
steepens the slope on the outside of a meander (figure
84). Once the ground ice is exposed it begins to melt
back rapidly, developing a semi-circular hollow opening
towards the river. The silt released by this melting
flows rapidly out of the central part of the hollow as a
mudflow tongue, flowing at a rate of between 5 m/sec and

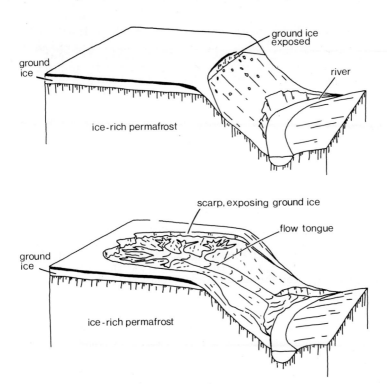

Figure 84   Block diagrams showing development of a bimodal
flow (modified from Lamothe and St. Onge 1961).

10 m/sec, but as it loses water it becomes more viscous,
moving more slowly down to the river as a series of
overlapping lobes (figure 84).   The authors suggest that
as the size of the hollow approaches 40 m in diameter the
ground ice becomes reburied by silt slipping down over it
and accumulating, melting then ceases and the mudflow
stabilises due to lack of water supply.   French (1976)
suggests that the majority of bimodal flows become
stabilised 30-50 summers after their initiation.

Detailed descriptions of the mechanisms of headscarp
retreat are given by Mackay (1966), Kerfoot (1969), and
McRoberts and Morgenstern (1974).   The main processes
include falling and sliding of the active layer over the
retreating ice scarp below;   simple ablation of the
exposed ground ice releasing soil which flows in pulses
down over the ice face to the mudflow lobe;   and where
the permafrost is not sufficiently ice-rich to sustain a
continuously ablating face, thawing of the scarp may
proceed to some critical depth, when failure occurs and
the thawed layer slides down to the mudflow, exposing the
still frozen material below to further thawing.   Rates of
scarp retreat are rapid, ranging from an average of 1 cm
per day to 20 cm per day during the ablation period
(table 5, p. 464, McRoberts and Morgenstern 1974).

Mackay (1966) shows that in very ice rich permafrost melting and retreat of the scarp may release a relatively small volume of mineral material so that scarp retreat can continue and the hollow grow even with no transport away of sediment. The floor of the hollow would then tend to be flat, with an extremely low gradient.

Although a southerly aspect would appear to favour rapid ablation and bimodal flow development it has been shown that they develop on slopes with all aspects (e.g. Kerfoot 1969). However, the fact that the sun does not set during the arctic summer means that steep slopes of any aspect may receive insolation.

Slab slides

Since slab slides on gentle slopes have not been described in the modern periglacial zone, except possibly in the form of coalescing active layer glides' but occur as fossil periglacial slope failures in temperate areas (noteably in southern England) they will be discussed in the next chapter, which deals with fossil periglacial slope deposits.

7. SLOPE DEPOSITS DUE TO PLEISTOCENE MASS-WASTING: THE EXAMPLE OF BRITAIN

Clearly information on mass-wasting processes and the properties of the sediments affected by such processes in the modern periglacial zone may be applied to the interpretation of fossil Pleistocene slope deposits elsewhere. To review world-wide Pleistocene deposits is beyond the space limitations of this book, and attention is therefore focussed on the British example. Sedimentological properties of the deposits are particularly stressed inorder that comparisons may be made with the modern periglacial sediments described in earlier chapters of this book. Only passing mention is made of the stratigraphical significance of these slope deposits in the Pleistocene chronology of Britain.

In this chapter slope deposits which have accumulated as a result of mass-wasting processes will be termed 'head', while finer-grained well sorted slope deposits resulting from the action of running water will be termed 'colluvium'.

Slope deposits of the Tertiary clays

Slab slides have been described in clays of Eocene, Cretaceous and Liassic age in Southern England and the East Midlands. They consist of shallow layers of head often containing rock fragments in a clay or silty clay matrix and underlain by roughly planar shear surfaces which lie approximately parallel to the ground surface and may extend continuously over considerable areas. In

Table 21   Summary of slab slide data from Kent and
            Oxfordshire (Weeks 1969)

Location	Slope	Depth of slip surfaces m	Depth m	Slope Deposit	Parent Material	c'_r kN/m^2	$\emptyset'_r$
Sevenoaks Kent	4^0	1.8-3.3	0-1.8	Chert and ragstone fragments in silty clay matrix	Weald Clay	1.4	15^0
			1.8-3.3	Finely fissured clay			
Tonbridge Kent (Lower Street)	4^0	0.9-3.0	0-0.9	Limestone fragments in silty clay	Weald Clay	2.1	16^0
			0.9-3.0	Fissured silty clay			
Tonbridge Kent (Quarry Hill)	7^0	1.05-4.5	0-1.05	Finely fissured silty clay with siltstone fragments	Wadhurst Clay	0	12.4^0
			1.05-4.5	Fissured clay			
Ditton Kent	3^0	1.2-1.8	0-1.8	Silty clay with flint fragments	Gault Clay	0	12.4^0
Broughton Kent	5^0	1.36	0-1.3	Slightly clayey sand with many flint fragments	London Clay	0	14^0
				Slip surface in upper few cm of weathered London Clay			

149

Table 21 (continued)

Location	Slope	Depth of slip surfaces m	Depth m	Slope Deposit	Parent Material	c'_r kN/m²	$\emptyset'_r$
Tetsworth Oxfordshire	3.5°	1.7-2.3	0-3.3	Silty clay with flint fragments	Gault Clay	0	14°
			3.3	Fissured clay			

many examples failure has occurred on gentle slopes well
below the limiting angle of failure under normal
conditions, but elsewhere slides on steeper slopes are
still active today.

Weeks (1969) provides the first detailed account of
slope deposits on gradients as low as 3° beneath which
slip surfaces indicate former instability. Field
observations were part of site investigation work for new
road schemes, and involved trial pits and bore holes.
Results are summarized in table 21.

Weeks analysed slope stability by applying the
infinite slope analysis of Skempton and DeLory (1957),
assuming seepage parallel to the surface. This shows that
while the 7° slopes are close to their limit of
stability, the 3° and 4° slopes would require pore
pressures equivalent to 0.54Z and 0.31Z respectively,
where Z is the depth of the slip surface. These pore
pressures are equivalent to r_u values of 0.72 and 0.65.
He concludes that such pore pressures could only have
developed under periglacial conditions.

The area to the south of the Lower Greensand
escarpment near Sevenoaks, Kent was subsequently
investigated in more detail by Skempton and Weeks (1976).
The oldest head deposits in this area occur as dissected
remnants of previously extensive sheets extending at a
gradient of about 1.5° at least 2 km from the escarpment
(figure 85). These are considered to be of Wolstonian
age. Following their deposition considerable erosion
occurred prior to the deposition of further periglacial
slope material, which extended up to 1 km to the south of
the escarpment. These deposits are of Devensian age, and
within some 300 m from the escarpment have been overrun
by a third sheet of head, which is clearly defined
morphologically by lobate frontal banks (figure 86).

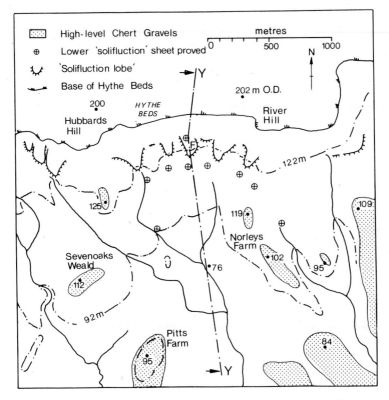

Legend (map):
- High-level Chert Gravels
- ⊕ Lower 'solifluction' sheet proved
- 'Solifluction lobe'
- Base of Hythe Beds

metres
0   500   1000

N

Figure 85   Map of the district around Sevenoaks Weald.
Section YY is shown in figure 86 (Skempton and
Weeks 1976).

An organic clay palaeosol was observed in the upper
surface of the main Devensian slope deposit, buried by
the uppermost sheet (figure 86), and gave an age of around
12000 radiocarbon years. This corresponds to the Allerød
Interstadial (Pollen Zone II), suggesting a Younger Dryas
(Pollen Zone III) age for the most recent periglacial
slope deposit.

The main Devensian slope deposit consists of clay with
embedded angular chert fragments, has a thickness of
around 2 m, and minimum gradient of slightly more than 2^0.
This overlies brecciated Weald Clay which contains
several slip surfaces in its uppermost layers. The upper
half metre or so of the main Devensian deposit consists
of silty clay slope wash material.

The Zone III periglacial slope deposits (referred to
as "solifluction lobes" by the authors) resemble the
earlier Devensian material, consisting of clay containing
chert fragments overlain by a surface slope wash deposit.
The general gradient of these deposits is 7^0. The clay
content of the matrix in Lobe F (figure 85) ranges from

151

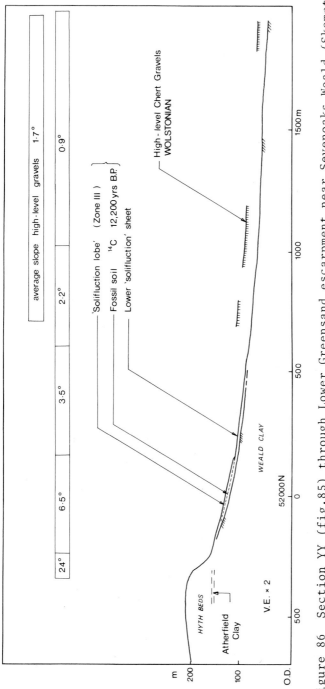

Figure 86   Section YY (fig.85) through Lower Greensand escarpment near Sevenoaks Weald (Skempton and Weeks 1976).

14% to 45%, with an average of 24%, while in the underlying main Devensian deposit the matrix contains 33% clay. The chert fragments generally range in size up to 20 cm in both sheets, but in the main Devensian deposit a fragment of length 60 cm was found. The plastic limit was 24 and 26 respectively for the Zone III material and the lower sheet, with plasticity indexes of 18 and 26.

Skempton and Weeks suggest that these deposits are best described as periglacial mudflows or mudslides, where the matrix-dominated clay gravel moved downslope principally by shearing on distinct slip surfaces. Stability analyses show that under temperate conditions, with the water table at the surface a gradient of 8^0 or more is necessary for such failures, but it is considered that under Pleistocene periglacial conditions seasonal thawing of an active layer above permafrost led to pore pressures sufficiently high to promote movement on slopes of less than 2^0. Assuming an active layer of 2 m, and that summer thawing took three months, the thaw consolidation ratio R has a value of 1.3, corresponding to an r_u of 0.89. Allowing R to vary by ±25%, to take account of the considerable uncertainties in its calculation gives a range of between 0.93 and 0.83 for r_u. Using measured strength parameters, limiting slopes were calculated corresponding to these periglacial conditions, and gave values of between 1.3^0 (r_u 0.93) and 3^0 (r_u 0.83). The authors also show that for failure on a slope of 2^0 the water content of the Weald clay would have been approximately 55%, and that of the matrix of the displaced material approximately 50%. These water contents are of course much higher than the present day values, and refer to conditions immediately following thaw.

Chandler (1970a) describes two localities in Northamptonshire where shallow slickensided slip surfaces approximately parallel to the ground surface and of considerable extent underlie relatively gentle slopes. At both sites rubble derived from the Northampton Sand (Inferior Oolite) set in a clayey sand or sandy clayey silt matrix forms a layer around 1 m thick overlying Upper Lias Clay. The superficial deposit becomes more clayey towards its base. At the more southerly site, Wellingborough, the surface gradient is 6.75^0, and at the second site, Isham, the slope is 4^0. In both cases the slip surface lies in the upper few cm of the Lias Clay. The clay at the slip surface is more plastic than the overlying coarser material (PI 37, Wellingborough, 27 Isham, compared with 4 to 21 in the overlying deposit). Shear strength parameters on the slip, of $c'_r = 0$ and $\emptyset'_r = 16^0$ indicate that under temperate conditions a slope of at least 12^0 is required to initiate movement. Chandler therefore concludes that the slip surfaces are relics of the conditions prevailing in a periglacial climate. He suggests that the high pore pressures necessary to cause movement over the slip surfaces (r_u 0.75) may have been

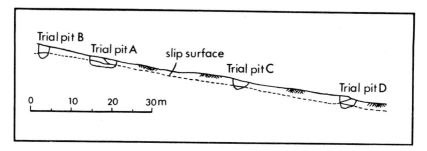

Figure 87   Section through slab slide in Lias clay, near Uppingham, Rutland (Chandler 1970b).

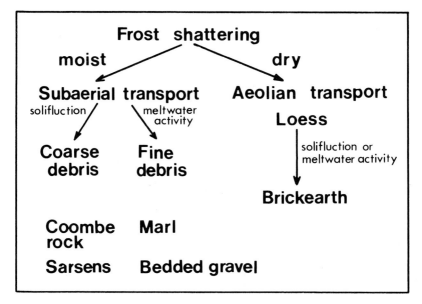

Figure 88   Periglacial deposits in Chalk (Evans 1968).

produced when water in the active layer became impounded on the slope between the permafrost table at the base and a shallow surface frozen layer due to frost. While such a mechanism may have operated, it would appear that thaw consolidation of ice rich Lias Clay could also have produced the high pore pressure necessary for shearing to occur.

Further north in the Upper Lias Clay outcrop Chandler (1970b) observed shallow slab slides similar to those described above, but entirely in clay. A polished planar slip surface at a depth of 1.5 to 2 m extensively underlies slopes of gradient around 9° (figure 87). Plastic limits of between 29% and 32% were measured in the disturbed clay, with plasticity index 28% to 44%. In this case the strength parameters at the slip surface are

$c'_r = 0$ and $\emptyset'_r = 18.5^0$, indicating that the slope is close to its limiting angle under present day conditions. In fact landsliding on part of the slope is currently active, although the slip surface extends beyond the limits of the modern slide (figure 87). Although these slopes are steep enough for movement today, the extent of the slip surface, and its obvious pre-dating of the current failure suggests again widespread shallow sliding on gentle slopes under Pleistocene periglacial conditions.

Finally, in the Swainswick Valley, near Bath, Somerset, Chandler et al. (1976) describe slopes of 9^0 with a 4 m thick blanket of head consisting of Great Oolite fragments in a clay matrix. The head overlies Fullers Earth (Jurassic age clay), and a continuous shear surface runs through its base. Recent instability over this slip surface is indicated by hummocky topography in places, but it again seems likely that it represents shallow sliding over permafrost during the Devensian. Index properties of the head show plastic limit 19% and plasticity index 28%, again relatively high compared with the sediments involved in modern periglacial mass movements.

The chalklands of southern England

Periglacial conditions during the Pleistocene cold phases led to the formation of dry valleys or coombes in the chalk, and solifluction was an important process of sediment displacement into them, the resulting deposit being known as coombe rock (Bromehead, in Dewey et al. 1924). Evans (1968) provides a useful summary of processes and deposits associated with periglacial weathering of chalk (figure 88). He describes coombe rock as heterogeneous material composed of coarse angular lumps of chalk set in a fine-grained matrix. Evans describes coombe rock in Wiltshire dry valleys up to 2 m in thickness, overlain by colluvial chalk muds. Kerney et al. (1964) show that much of the chalky drift in coombes in the North Downs escarpment consists of colluvium. Such an interpretation is supported by the mulluscan fauna present in these deposits. However, Kerney et al. (1964) note up to 1 m of chalk rubbles, muds, and clayey chalk gravels beneath the colluvium in the aprons of drift extending beyond the mouths of the dry valleys over the Gault Clay vale.

A palaeosol of Zone II (Allerød) age often divides the colluvium sequence, and may show signs of cryoturbation (Evans 1968). The coombe rock is therefore generally considered to be of Devensian age, the colluvium below the palaeosol of Late Devensian (Zone I) age, and the colluvium above the palaeosol to be Zone III (Younger Dryas) age.

Also incorporated in slope deposits over chalk may be brickearth, consisting of silt-sized quartz with numerous

155

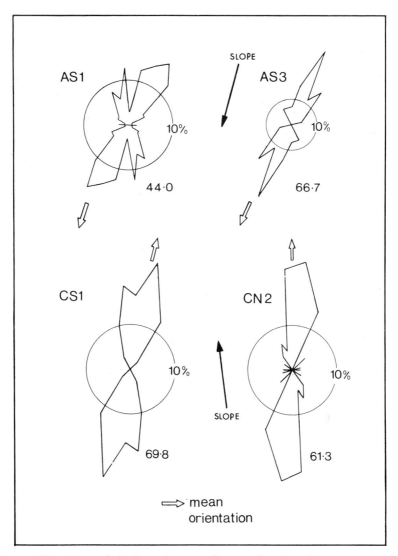

Figure 89   Macrofabrics in coombe rock, South Downs,
             Hampshire (Shakesby 1975).

chalk pellets (Evans 1968). This material is considered
to be of loessic origin, but to have suffered post-
depositional periglacial mass movement. Similarly the
clay-with-flints may have suffered periglacial mass
movement down valley sides, extending their distribution
from their original *in situ* nucleus.

        Small et al. (1970) consider that the sarcen stones
of the Wiltshire chalk were carried into the valleys as
part of solifluction material. At Clatford Bottom in the
Marlborough Downs up to 3.5 m of soliflucted coombe rock

mantles the lower parts of the gentle south-facing slope.
The upper 1.2 m has suffered post-depositional
decalcification, with consequent reduction in thickness,
darkening in colour and concentration of sarcens and
flints. The coombe rock consists of a white (10YR8/1) to
light grey (10YR7/2) chalk rubble set in a pasty or
powdery chalk matrix, and grades downwards into *in situ*
weathered chalk.

The gentle slopes over which this material has moved
(3^0 to 4^0) suggest the development of considerable pore
pressures during mass movement, due to the melting of ice
lenses and the release of supernatent water during thaw.
On 3^0-4^0 clay slopes mantled with periglacial slab slides
Weeks (1969) showed that r_u values of up to 0.72 would
have been necessary to generate mass movements, and it is
likely that the frictional strength of the fine chalk
matrix is higher than that of the weathered clays in the
Weald. French (1973) also stresses the low angle of
south-west facing slopes in asymmetrical chalk valleys.
These gentle slopes are mantled with up to 3 m of
solifluction deposit consisting of chalk rubble in a fine
grained calcareous matrix.

Detailed sedimentological data relating to coombe rock
in a dry valley near Petersfield in the South Downs of
Hampshire are given by Shakesby (1975). Two main
sedimentary units are identified near the valley floor,
an upper white (10YR3/2) matrix containing flints and
chalk fragments, and a lower buff-coloured matrix (10YR8/
2) containing marcasite nodules, ironstone fragments,
shattered flints and chalk fragments. Below about 2.2 m
the angular chalk clasts become increasingly large. The
upper white chalk matrix is slightly finer textured than
the lower buff-coloured material (Ømean -0.93 and -0.143
respectively) but both contain less than 12% silt/clay.
Both deposits are non-plastic, with plasticity index 3-6%
and liquid limit 26-30% for the upper material and
plasticity index 3%, and liquid limit 23% for the lower.
Long axis orientation measurements showed strongly
developed downslope orientation of elongate clasts,
(figure 89) and Shakesby concludes that solifluction was
the main agent of sediment movement into the valley.

South West England

The widespread occurrence of solifluction deposits or
head (De La Beche 1839) in South West England is well
illustrated in the Memoirs of the Geological Survey for
the region. For instance, Edmonds et al. (1968)
describing the Okehampton area of Devon state that
geological mapping is rendered difficult by head deposits
which are present almost everywhere, in thicknesses
commonly of 1.8 to 3 m. Similarly, in the area around
Boscastle and Holsworthy in North Cornwall and West Devon,
McKeown et al. (1973) observe that head is common
throughout the district, ranging in thickness from 0.9 to

3.7 m.

The nature of these deposits is related closely to
the underlying geology. McKeown et al. (1973) describe
bluish clays containing angular clasts overlying the
slaty and shaly facies of the Devonian and Carboniferous
rocks in north Cornwall and west Devon, while the more
arenaceous Bude formation gives yellowish-brown silts,
locally clayey or sandy and containing angular rock
fragments. Over the Permian outcrops is spread a cover of
brownish red gravelly silts, while the clayey head
overlying the adjoining Carboniferous rocks is commonly
reddened up to 300 yards (274 m) beyond the boundary
between the two rock types. Edmunds et al. (1968) show
that on the northern flanks of Dartmoor, granite head may
also extend down over the surrounding metamorphic rocks
of the aureole.

An early account of mass movement of granite head
beyond the granite outcrop is given by Hill and McAlister
(1906) in their description of the geology of the
Falmouth-Truro-Camborne area of west Cornwall. In a
railway cutting adjacent to the granite hill of Carn Brea
is exposed 'a deposit of granite detritus, several feet
in thickness, resting upon killas. In the finer material
blocks of granite are incorporated some of which attain a
size of 5 to 6 feet. The granite margin is probably about
70 yards distant at the west end of the section and about
180 yards at the eastern portion. If the ground were
steep, this deposit at such a short distance from the
granite would not be unusual, but as a matter of fact the
land here is comparatively flat, the angle of slope being
only 1^0 to 2^0, while the declivity from the base of Carn
Brea Hill extending to a distance of 500 yards only
averages 4^0. It is clear, therefore, that the large
granite blocks could only have reached their present
situation under abnormal conditions' (p.95). Such
"abnormal conditions" must have closely resembled the
Devensian permafrost climate described by Skempton and
Weeks (1976) to explain the low-angled mass movements in
the Weald Clay of Kent, and were probably synchronous with
them.

Waters (1964) describes two facies of head on
Dartmoor, where on the lower slopes weathered granite or
growan is overlain by a sandy head containing small
fragments of bedrock. Above this 'main head' is an upper
head containing much larger blocks which merge with the
clitter at the surface. Waters postulated two phases of
mass movement, with the upper layers of regolith
containing finer rock fragments being spread over the
lower slopes first, and the lower, coarser layers
suffering mass movement later and covering the slopes
with coarser head, giving an inverted sequence. Green
and Eden (1973) however found considerable variation in
the distribution of coarser granite blocks with depth,
and concluded that larger blocks are just as likely to be

concentrated at the base of the head as at the surface.
They therefore suggested that only one phase of
periglacial mass movement is indicated. In ten samples
taken from three sites Green and Eden found average
textural properties for the granite head matrix of 52.6%
sand, 34.5% silt and 12.9% clay. Matrix material such as
this falls well within the range of texture observed in
soliflucted sediments of the modern periglacial zone
(figure 56).

The head deposits of South-West England are probably
best known from coastal exposures. Deposits may be of
considerable thickness; up to 14.4 m at Saunton, north
Devon (Kidson 1971) and up to 35 m in the Valley of Rocks
near Lynmouth (Palzell and Durrance 1980). Two distinct
units are often recognized, a lower "main head", which is
generally thickest, consisting of angular blocks of local
rocks set in a sandy matrix (Stephens 1970), and a more
variable "upper head". The upper head may rest directly
on the main head (e.g. Trebetheric Point, Arkell 1943) or
be separated from it by a more sandy layer (e.g.
Middleborough, Croyde Bay, figure 11.4, p. 280 Stephens
1970), or by sand rock (e.g. Widemouth Bay, North
Cornwall). In many of the small coastal valleys of North
Cornwall and North Devon only one unit of head is present,
which may be over 4 m in thickness. In South Devon, at
Wembury Bay, Kirby (1967) reports only one unit of head
up to 9.1 m in thickness, consisting of angular fragments
of the underlying Dartmouth Slates set in a sandy matrix,
although Harris (in preparation) has observed
discontinuous layers of colluvial silt within this head.
Further east on the South Devon coast between Start
Point and Hope Cove Mottershead (1971) describes up to
three units of head differentiated largely on the basis of
texture.

As mentioned above, textural properties of the head
deposits depend largely on the nature of the bedrock, but
Kidson (1971) has pointed out that variation in the
relative importance of solifluction and slope wash may
lead to variation in the overall textural range.
Mottershead (1971) stresses the coarseness of the head
deposits of the South Devon coast. He observed maximum
stone sizes exceeding 260 mm, and suggested that the
closeness of jointing in the bedrock will largely control
the size of frost weathered debris. Stone contents ranged
from 20% to 90% with a mean of 54%, and all samples fell
within poorly sorted or extremely poorly sorted
categories. It should be emphasised, however, that there
may be considerable textural variation within the head at
one site, particularly with respect to the relative
abundance of stones and matrix. This is well illustrated
in the detailed study by E. and S. Watson (1970) of the
coastal head deposits in the Cotentin Peninsula,
Normandy, deposits which closely parallel the periglacial
slope deposits in south west England.

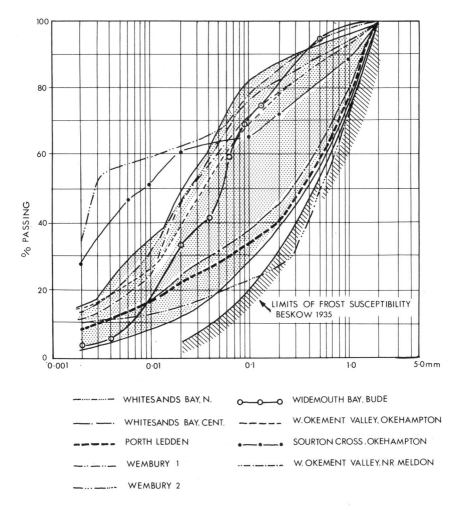

—·····—·— WHITESANDS BAY, N.	o—o—o WIDEMOUTH BAY, BUDE
—·—·—· WHITESANDS BAY, CENT.	—————— W. OKEMENT VALLEY, OKEHAMPTON
————— PORTH LEDDEN	•—·—•—• SOURTON CROSS, OKEHAMPTON
—··—··— WEMBURY 1	··—·—··—·· W. OKEMENT VALLEY, NR MELDON
—···—····— WEMBURY 2	

Figure 90 Textural curves for finer than 2 mm fractions
from selected sites in S.W. England.
Whitesands Bay and Porth Ledden from Watson
and Watson 1970, Wembury from Harris in prep.,
Okehampton sites, data kindly supplied by J.
Harris of F.W. Sherrell (Consulting
Geologists) Tavistock, and shaded area,
textural envelope for coastal head deposits,
Start Point to Hope Cove, Mottershead 1971.

Since sediments accummulated by mass-wasting processes
must necessarily behave as matrix-dominated soils during
accumulation, textural and strength characteristics of
the finer fractions are of considerable importance.
Mottershead (1971) compared the finer than 2 mm fraction
from head in South Devon with the frost susceptibility
limits of Beskow (1935) and Cailleux and Taylor (1954).
The matrix was found to be frost susceptible, suggesting

Table 22   Index properties of head deposits in south west England

Location	Liquid Limit %	Plastic Limit %	Plasticity Index %	% Clay	Author
Start Point to Hope Cove, S. Devon. (main head)	29	19	10	6	Mottershead (1970)
	26	19	7	14	
	26	17	9	14	
	28	17	11	9	
	21	17	4	8	
	26	17	9	3	
Wembury Bay, S. Devon. (main head)	27	19	8	10	Harris (in prep.)
Wembury Bay, S. Devon, (colluvial layer in head)	35	23	12	35	Harris (in prep.)
Widemouth Bay, Bude, N. Cornwall, (main head)	23	16	7	4	Harris (in prep.)

that ice segregation was likely to have occurred during freezing (figure 90). Similarly, the matrix material in head elsewhere in South West England is often in the frost-susceptible textural range (figure 90). The finest grained material in figure 90 is from a silty layer some 0.5 m thick near the base of the head exposed in Wembury Bay on the south coast of Devon, interpreted as a colluvial layer within the head. Here clay content is 35% of the finer than 2 mm fraction, but in all but one of the other samples, clay content is less than 15%. The exceptional case, with 28% clay content is derived from the baked carboniferous rocks of the Dartmoor aureole, and is from the Okehampton area, in West Devon.

Index properties where reported generally indicate a low plasticity matrix, with plasticity indices in the range of 4 to 11% (table 22). Such low plasticity indices suggest that high pore pressures generated during thaw-consolidation would lead to flow-dominated slope failure (solifluction, mudflow) rather than the slip-dominated failure in the clays of south-eastern England.

The organisation of their fabric elements strongly supports the suggestion that these head deposits accumulated by solifluction. It is widely reported that elongate stones show a strong preferred orientation parallel to the direction of slope (figure 91) and low-angled dip often roughly parallel to the ground surface (Stephens 1961, Kirby 1967, Kidson 1971, Mottershead

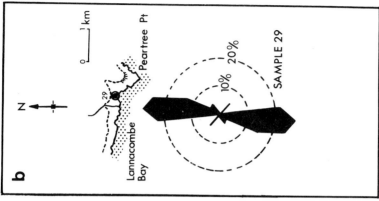

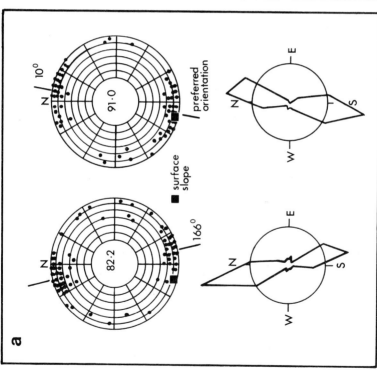

Figure 91 Macrofabrics from coastal head deposits, S. Devon, (a) Wembury (Kirby 1967), (b) Peartree Point (Mottershead 1971).

Figure 92   Vertical thin section of undisturbed head
             material from Wembury S. Devon, showing low-
             angled dip of sand and gravel sized fragments.
             Surface slope 5° to the left.

1971). The tendency for platy fragments to dip at a low
angle produces a crudely bedded appearance in many
instances. The organisation of fabric elements extends
down to the sand-sized particles (Harris, in
preparation). Figure 92 shows a vertical thin-section of
undisturbed head from Wembury, South Devon in which the
low angle dip of sand and gravel grains is clearly
illustrated.

On the basis of two head deposits recognised in North
Devon, Mitchell (1960) and Stephens (1961) date the main
head as Wolstonian and the upper head as Devensian.
Kidson (1971) and Mottershead (1977) however consider
that only the Devensian period is represented in the heads
of south-west England. They maintain that textural
changes in the head are related to variations in the
bedrock, inputs of colluvial silts due to climatic
variation, and in coastal sites, inputs of blown sand
derived from the underlying raised beaches.

Wales

Unlike southern England, Wales was glaciated during
the Wolstonian, and apart from small areas in South Wales,
was again subject to glaciation during the Devensian.
Periglacial mass-wasting processes operating during the
non-glacial phases of these cold periods therefore
affected a greater range of deposits in Wales than was
the case in southern England. Where head forms the basal
member of Pleistocene deposits, such as in the coastal
exposures of northern Pembrokeshire (John 1970) it may
consist of angular fragments of bedrock in a finer

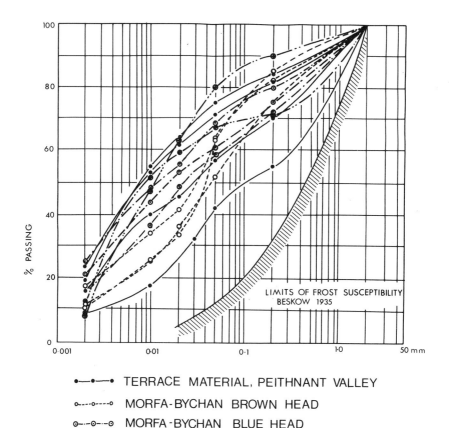

•—•—•	TERRACE MATERIAL, PEITHNANT VALLEY
o----o----o	MORFA-BYCHAN BROWN HEAD
⊙--⊙--⊙	MORFA-BYCHAN BLUE HEAD
⊕···⊕···⊕	MORFA-BYCHAN YELLOW HEAD

Figure 93   Textural curves for finer than 2mm fractions
from terrace material, central Wales and
Morfa-bychan coastal exposures (data from
Watson 1976).

matrix, merge downwards into overturned frost weathered
bedrock, and closely resemble the head deposits of South
West England. Elsewhere head deposits often incorporate
older glacial drifts (e.g. the heads of the Gower coast,
Bowen 1970), or may consist largely of reworked till (e.g.
the soliflucted Devensian till of the South Wales
coalfield, Harris and Wright 1980). In some cases
considerable doubt prevails as to the precise origin of
soliflucted deposits, for instance at Morfa bychan, south
of Aberystwyth.

South and Central Wales

    The occurrence of solifluction terraces in valley
bottoms in the Radnor Forest, Mynydd Bach, Mynydd Eppynt
and Fforest Fawr-Brecon Beacons areas of central and

164

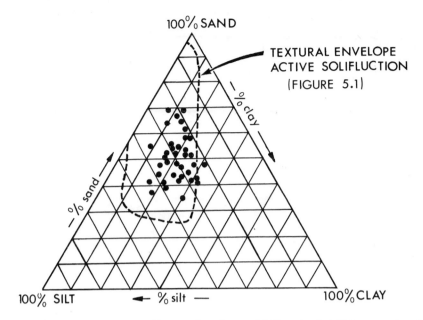

100% SAND

TEXTURAL ENVELOPE
ACTIVE SOLIFLUCTION
(FIGURE 5.1)

% clay

% sand

100% SILT ← % silt — **100% CLAY**

Figure 94   Texture of matrix in soliflucted till, central
Wales (Potts 1971).

south-central Wales has been reported by Crampton and
Taylor (1967), and Crampton (1965) has shown that much of
the terrace material is indurated by densely packed silt
and clay, a condition considered to be the result of
permafrost and ice segregation. Lewis (1970) describes
similarly widespread solifluction deposits on hillsides
and in valley bottoms in the Black Mountains and Brecon
Beacons. In the uplands drained by the Dovey, Rheidol and
Ystwyth, west-central Wales, Watson (1970, 1976)
describes smooth solifluction terraces sloping down to the
river, developed over bedrock which breaks down to
produce silt-rich detrital material. The matrix in these
deposits is frost-susceptible (figure 93) and generally
texturally similar to the heads in S.W. England described
earlier in this chapter. Such terraces do not occur over
bedrock which weathers into joint blocks, with few fines.
In mid Wales, nivation hollows also show terraces and
fans of solifluction deposits extending down towards the
valley bottoms (Watson 1966). These fans have surface
slopes as low as 4°. Crampton and Taylor suggest that
terraces occur at the base of slopes of all aspects, but
are generally best developed on slopes facing south,
while Watson's research indicates that they occur most
frequently at the foot of north and east facing slopes.

Watson (1970) considers that the close relationship
between bedrock and the distribution of solifluction
deposits, and the fact that generally the higher the slope
to the rear of the terrace, the thicker the accumulation,
indicate that these slope deposits result from

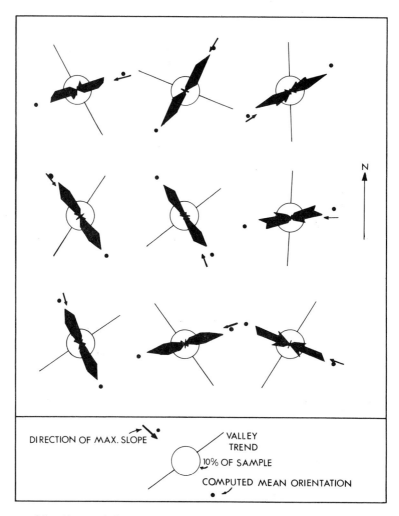

Figure 95 Macrofabrics in soliflucted till, central Wales
(Potts 1971).

mass-wasting of the products of frost weathering, and
are   therefore entirely the result of periglacial
conditions. Potts (1971), however, suggests that the
relatively high percentage of fines in terrace material
from central Wales precludes a frost shatter origin for
these drifts, and he concludes that the terraces consist
of local till reworked by solifluction. The silt/clay
content of the terraces is generally in excess of 40%
(figure 94), and the stones in the terraces are sub-rounded
to sub-angular and heavily striated. The orientation of
clasts in these terraces confirms their accumulation by
downslope mass movement, with strongly developed preferred
orientations parallel to the direction of slope (figure
95). Potts also examined the distribution of terraces

with respect to aspect and found that although in broad,
gently sloping valleys soliflucted till mantles slopes on
both sides of the river, steep-sided deeply incised
valleys tend to show terraces on one side of the river
only. However, statistical analysis of the aspects
observed shows no preferred aspect for the development of
solifluction, and Potts concluded that the distribution of
till prior to its reworking largely controlled the present
day distribution of solifluction terraces in the valley
bottoms.

In the South Wales coalfield head mantles hillslopes,
and merges with till (Bowen 1970). Harris and Wright
(1980) describe up to 6 m of head overlying lodgement till
on the valley side of a headstream of the River Ely, near
Llantrisant. The head consists of locally derived
angular and sub-angular pebbles and gravel in a silty sand
matrix, the latter containing 57% sand, 36% silt and 7%
clay, compared with 32% sand, 61% silt and 7% clay in
the lodgement till below. The underlying till contains
locally derived clasts, with a small input of Old Red
Sandstone erratics. No significant difference was
observed in Cailleux Flatness indices between the
solifluction deposits and the till (240-305, solifluction
deposit, 237-315, till) but the solifluction material was
slightly more angular (Cailleux Roundness 45-52
solifluction deposit, 73-82 till), the difference being
significant at the 99% level. The presence of subangular
striated clasts in the solifluction deposit and the
similarity in lithologies between the till and the
overlying material led Harris and Wright to conclude that
the solifluction deposit consists of reworked till with
the addition of frost-shattered bedrock derived from the
hill summit.

The main discriminating factor between the
solifluction deposit and the till at this site is fabric,
with the lodgement till showing preferred orientations
parallel to the valley trend and the overlying
solifluction deposit showing strong preferred orientation
down the valley side, parallel to the slope and
perpendicular to the stone orientations in the till
(figure 96). In the solifluction deposit the tendency for
platy fragments to dip parallel to the surface, and the
presence of discontinuous silty layers gives the deposit
a crudely bedded appearance in section.

The surface slope on this valley side is only 6^0 to 7^0,
and on the basis of data from similar deposits elsewhere
in the South Wales coalfield a value of $\emptyset$ of around 30^0
might be anticipated for the soliflucted material (Wright,
personal communication). Since the matrix material falls
in the frost susceptible range of textures, the thaw-
consolidation process probably operated here during the
periglacial phase of head accumulation, generating the
high pore pressures necessary to promote instability on
such a gentle slope. Index properties indicate that both

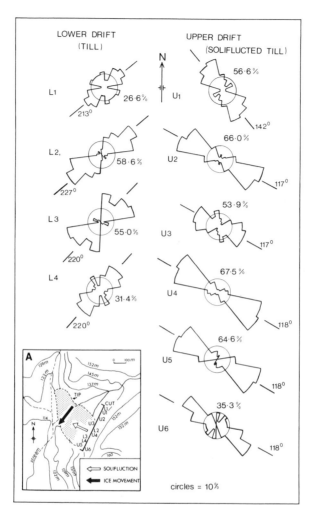

Figure 96   (a) Macrofabrics from lodgement till and overlying solifluction deposit near Llantrisant, S. Wales (modified from Harris and Wright 1980).

the till and solifluction material are non-plastic (LL 26% and 30% for head and till respectively, PI 10% and 14%) indicating that solifluction was the likely mode of mass movement rather than slip failure. No slip surfaces were observed in the soliflucted material, or between it and the underlying till. The till is of Devensian age at this site, the phase of periglacial mass-wasting must therefore date from Late Glacial times, Zone I and III.

On slopes in the coalfield with only a thin till cover, frost shatter of the bedrock provided regolith which was subsequently moved downslope by mass-wasting

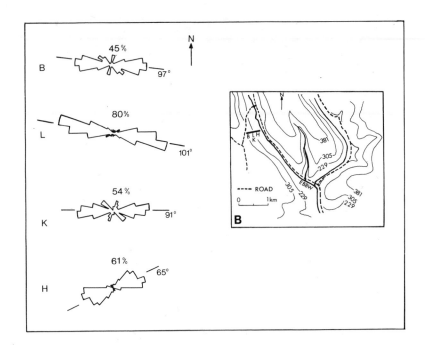

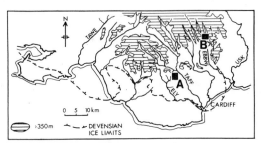

Figure 96   (b) Macrofabrics in head on the western side
of the Ebbw Valley near Ebbw Vale, S. Wales
(data kindly supplied by M.D. Wright).
Figure 97 illustrates the section from which
these fabric diagrams were obtained.

processes. On a steep hillside near Ebbw Vale, Wright
and Harris (1978) and Harris (1980) describe such a site.
Frost shattered sandstone fragments incorporated in the
head may be traced back to severely disrupted bedrock
(figure 97). Clasts are angular and platy and are set in
a silty sand matrix. Orientation and dip measurements
made by M.D. Wright (Wright, in preparation) clearly
demonstrate fabric organisation strongly related to the
slope (figure 96). Harris (1980) has examined the
micromorphology of this material and found silt cappings
covering the finer clasts which closely resemble cappings

Figure 97   Exposure in head on the western side of the
Ebbw Valley near Ebbw Vale, S. Wales.

observed coating sand grains in active solifluction
material from North Norway (chapter 5, page 137).
Similar micromorphological features were also observed in
thin sections made from solifluction material at Taf
Fechan in the Brecon Beacons.

On the South Wales coast the extensive exposures of
head in the southern cliffs of Gower have been described
by Bowen (1970).   Here the clastic material includes
erratics derived from an earlier till.   Fabric
measurements confirm the slope-controlled origin of these
deposits (Harris 1973).   This area lay beyond, but close
to the Devensian ice limits (Bowen 1970), so that no
Devensian till is incorporated in the heads.   Head
deposits form a smoothly concave apron infront of a
fossil cliff line at Rhossili, between Port Eynon and
Oxwich, and at Heatherslade and Hunts Bay.

Cardigan Bay and North Wales

Coastal deposits in Cardigan Bay include head and
Irish Sea till.   Watson and Watson (1967) provide a
detailed description of the head material at Morfa-bychan,
south of Aberystwyth, where three facies of head
consisting of material derived from the Aberystwyth Grits
are exposed.   The Aberystwyth Grits (siltstones,
sandstones and cleaved mudstones) are the only rocks
outcropping in this coastline and for some 7 km inland
(Watson 1976).   The basal member of the sequence
consists of a yellow-coloured deposit containing
weathered angular rock fragments in a yellowish-grey silt
matrix.   The clasts dip away from the cliff at angles of
30^0 to 40^0, giving a crudely stratified appearance to the
deposit.   The clasts are strongly oriented

170

perpendicular to the cliff face, parallel to the slope behind the cliff, with 76% to 84% of clasts within 15^0 of the median.

Overlying this material is a blue-coloured head with a silty matrix, and containing clasts up to 1 m or more. The clasts are less angular than in the yellow head below, with worn surfaces and edge-rounding. They show similar preferred orientations to the clasts in the yellow head, but a greater range, with only 60% to 76% of the clasts within 15^0 of the median. Bedding dips away from the cliff face at between 15^0 and 20^0, but the clasts show slight imbrication relative to bedding. Associated with the head are thin beds of washed fine gravel, sand and occasionally laminated silt, interpreted as slope wash deposits. Watson and Watson (1967) suggest that some of the abraded and striated clasts may represent older (pre-Devensian) till material incorporated in the head by Devensian solifluction, but that most of the deposit is the product of frost shattering and solifluction of the bedrock in the coastal slope behind the exposure.

The uppermost deposit exposed at Morfa bychan is a brown head formed of angular unweathered bedrock fragments in a fine matrix. The clasts are very variable in size and quantity, and the matrix is lower in clay and silt than in the underlying deposits. In all three head facies the matrix material falls in the frost-susceptible textural range (figure 93), and there is no doubt that they were deposited in their present location as a result of periglacial slope processes.

In the uplands of North Wales till has been extensively reworked by solifluction (Whittow and Ball 1970). Ball et al. (1969) describe smoothly sloping soliflucted Devensian till mantling the sides of the valleys south-east of Snowdon. Mechanical analysis of the drift matrix show mean values of 4% clay, 24% silt and 72% sand. The periglacial mass-movements which reworked these deposits are considered to have taken place during Zone III times when the corries on Snowdon were reoccupied. In the Rhinog Mountains, Ball and Goodier (1969) describe large sorted stripes, the surface expression of late glacial frost sorting and solifluction, and in Snowdonia, Ball and Goodier (1970) report valley-side solifluction terraces, turf- and stone-banked solifluction lobes, and sorted and unsorted stripes. These are considered to be of Late Glacial age, but some of the smaller lobes and stripes may be more recent, and some may be active today.

Northern England and Scotland

Tufnel (1969) has reviewed the literature on periglacial features in northern England, and stressed that the role of periglacial slope processes has been neglected. Solifluction deposits in the Lake District, Howgill Fells, the Pennines, the North York Moors and the Cheviots are mentioned by Tufnel. Hollingworth's paper on periglacial phenomena in the Lake District (Hollingworth 1934) is well known. He describes stone stripes, and solifluction lobes, and provides an excellent photograph of a small stony solifluction lobe. Dines et al. (1940) describe the occurrence of head in valley bottoms and on the lower slopes in West Yorkshire. Here the deposit varies widely in composition according to the nature of the parent material (solid or drift) on the slopes above. In Millstone Grit and Coal Measures areas it is generally an unbedded loam or clay containing abundant angular rock fragments. In places stones derived from boulder clay are present, and these are generally more rounded. In many areas the head consists of reworked till. Dines et al. suggest that on the basis of the distribution of head deposits, solifluction processes were more active on north- and east-facing slopes than on south- and west-facing slopes. Eden et al. (1957) also describe extensive head deposits on the Millstone Grit and lowest Coal Measures to the west of Sheffield, in the Pennines. The deposits are clayey, sandy or gravelly loams containing abundant blocks of sandstone. The matrix is derived from shale while the blocks are the product of weathering of the Millstone Grit escarpment. Eden et al. (1957) observe that the deposits are clearly the product of flow, and are thin or absent on spurs, but thick in depressions. Movement has occurred on slopes as low as 3°, indicating high water contents and the development of considerable pore pressures.

King (1976) and Thomas (1977) suggest that the subdued forms of most of the high ground in the Isle of Man indicates that solifluction was active over a long period. Thomas (1977) describes the thickening of local drift deposits downslope to form drift terraces which are generally better developed below the north-facing valley sides. This produces a marked asymmetry in the drift distribution of the upland valleys in the Isle of Man. On interfluves and summits the drift is a thin angular rubble, but on lower slopes the terraces comprise more compact silty head often showing rough stratification parallel to the slope. The fabric is everywhere strongly developed, with elongated clasts orientated downslope. Subdividing the matrix-dominated head into three units are slope wash gravels, suggesting three cycles of solifluction followed by slope wash. The final slope wash phase was followed by the accumulation of upland peat and soil formation.

In Scotland, Galloway (1961) indicates that solifluction deposits cover many of the hillsides. Till has often been reworked to a depth of at least half a metre by periglacial mass-wasting, the distinction between soliflucted and non-soliflucted till being based

largely on stone orientations. The nature of the
solifluction deposit depends on the nature of the till,
or in areas lacking till cover, the bedrock. In most of
north-eastern Scotland and the eastern Southern Uplands
frost shattering has produced head with a silty matrix
containing angular stones. The head forms smooth sheets
usually 0.5 to 1 m in thickness, but thicker in hollows
and in many valley bottoms. In the Highlands,
solifluction material is generally bouldery. Here the
topographic features of mass-wasting, stone-banked
terraces and lobes, have often suffered loss of fines and
are consequently largely composed of stones and boulders.

Detailed studies of the head deposits in the
Southern Uplands have been made by Tivy (1962) and Ragg
and Bibby (1966). Tivy examined the Lowther Hills, and
Ragg and Bibby studied Broad Law, in the Central Southern
Uplands. In both cases a superficial gravelly layer was
found, lacking in fines, and consisting of flat-lying
angular platy fragments with strong preferred
orientations parallel to the slope. Beneath this was
a poorly sorted deposit with angular blocks set in a
silty or fine sand matrix. The blocks again showed
preferred orientations parallel to the slope, though less
strongly developed than in the gravels above, and Ragg
and Bibby report a marked tendency for the stones to dip
into the slope in an imbricate fashion. In both studies
matrix material in the poorly sorted deposit was observed
forming compact coatings on the upper surfaces of clasts,
while the lower surfaces were clean and smooth. These
may be of similar origin to the matrix coatings described
by Harris (1980) in South Wales. Tivy concludes that the
absence of fines in the upper layer and its rudimentary
stratification indicates a grèze littées or stratified
scree, but considers the origin of the lower material to
be conjectural. Ragg and Bibby however interpret both
deposits as the product of one weathering-mass movement
period. The upper gravelly material is considered to
have lost fines through downwashing by snowmelt, or to be
the result of the concentration of coarser material at
the surface by frost sorting. Solifluction in the lower
matrix-controlled material is considered to be the most
important mechanism of sediment displacement down
hillsides and into the valley bottoms.

Further north, on the western side of Loch Lomond,
Rose (1980) describes smooth slopes mantled with
soliflucted Devensian till. Frost-susceptible silt-rich
tills on the north side of Glen Fruin form typical
solifluction slopes, with a gradual thickening of the
deposit towards the valley bottom. Fabric analysis shows
strong preferred orientations of clasts parallel to the
slope. Elsewhere in the area, where till lacks a high
silt content, extensive reworking by solifluction has not
occurred. Since soliflucted till occurs outside the ice
limits of the Loch Lomond readvance and not inside,
Sissons (1977) and Rose (1980) consider that this phase

of periglacial mass-wasting occurred during the Loch Lomond stadial, that is, in Pollen Zone III.

More recent periglacial mass movements are reported in the Highlands, where in the Cairngorms, King (1972) suggests some small turf-banked lobes may be active today, although the larger stone-banked terraces and lobes are considered to be relict. Ballentine (personal communication 1980) has shown that turf-banked lobes on steep slopes on Ben Wyvis in Easter Ross are presently active, and FitzPatrick (1958) observed that mass movements of frost shattered debris are active in the Cairngorms and in the western and southern Highlands. Sugden (1971) suggests that solifluction was active in the Cairngorms during the 'Little Ice Age' although he considers its effect on the landscape to have been only minor. He quotes radiocarbon dates from organic material buried beneath solifluction lobes of 4880 ± 135 and 2680 ± 120 years BP, indicating that solifluction has certainly been active since these dates. White and Mottershead (1972) found organic material with a radiocarbon age of 5145 ± 135 BP beneath a solifluction terrace on Ben Arkle, in Sutherland, again indicating active solifluction at some time between that date and the present.

Summary and concluding comments

Two distinct types of deposit resulting from periglacial mass-wasting have been described in Britain. The clays of southern England have suffered shallow landsliding on low-angled slopes. The base of the displaced sediment is therefore formed by one or more slip surfaces and the displaced mass may suffer varying degrees of disturbance. These clay heads show relatively high plasticity with plasticity index ranging from 20% to 40% or more. In contrast, the head deposits resulting from periglacial mass movement of frost shattered bedrock and those resulting from mass movement of till, generally contain much smaller amounts of clay, the matrix usually consisting of silt and fine sand. These deposits contain rock fragments orientated in the direction of slope, indicating that movement was by a flow-type of displacement (gelifluction or frost creep) rather than by slip failure. No shear surfaces are reported from these deposits. Lack of clay leads to a non-cohesive and non-plastic matrix with plasticity index generally less than 15% and liquid limit less than 30%. Comparison of texture and index properties shows general similarity between these deposits and modern solifluction sediments (figure 98). It is noteworthy that the index properties for the clay heads of southern England are distinctly higher than those of modern solifluction sediments.

Process studies of mass-wasting in the modern periglacial zone, coupled with detailed sedimentological and geotechnical analyses are necessary to furnish the

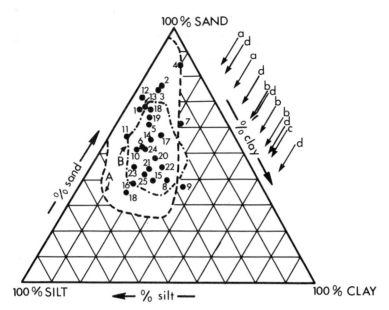

Figure 98   Textural properties of the finer than 2mm
fractions of solifluction and other
periglacial mass-wasting deposits in Britain.
A Envelope of modern solifluction sediments
   (figure 56).
B Envelope for soliflucted till, central Wales.
1, Widemouth Bay, N. Cornwall.
2, Porth Ledden, S. Cornwall.
3, Whitesands Bay (central), S. Cornwall.
4, Wembury, S. Devon.
5, W. Okement Valley, Okehampton, W. Devon.
6, Whitesands Bay (northern), S. Cornwall.
7, W. Okement Valley, Okehampton, W. Devon.
8, Sourton Cross, Okehampton, W. Devon.
9, Wembury, S. Devon, colluvial layer.
10, Dartmoor, Devon.
11, Ely headstreams, S. Wales.
12, Snowdon, soliflucted till, N. Wales.
13, 14, 15, 16, Peithnant Valley, central Wales.
17, 18, 19, Brown Head, Morfa-bychan, Cardigan
    Bay, Wales.
20, 21, 22, 23, 24, Blue Head, Morfa-bychan,
    Cardigan Bay, Wales.
25, Yellow Head, Morfa-bychan, Cardigan Bay,
    Wales.
Clay contents;
 a, Shallow slides in Weald Clay, Sevenoaks,
    Kent.
 b, Shallow slide over Lias Clay, Uppingham,
    Rutland.
 c, Fullers Earth in which shallow slides occur,
    near Bath, Somerset.
 d, Shallow slides developed over Fullers Earth,
    near Bath, Somerset.

information required for the correct interpretation of
fossil periglacial slope deposits. When we can show that
certain periglacial environments act on sediments with
certain sedimentological and geotechnical properties to
produce a certain range of processes we may then be able
to reconstruct with confidence the Pleistocene periglacial
environments which prevailed in Britain and produced the
deposits described in this chapter.

8. REFERENCES

Aldrich, H.P., 1956. Frost penetration below highway and
    airfield pavements. *Highway Research Board Bulletin,*
    135, 124-149

Alexander, C.S. and Price, L.W., 1980. Radiocarbon
    dating of the rate of movement of two solifluction
    lobes in the Ruby Range, Yukon Territory.
    Quarternary Research, 13, 365-379

Andersland, O.B. and Anderson, D.M., 1978. *Geotechnical
    Engineering for Cold Regions,* (McGraw-Hill, New
    York)

Andersson, R.G., 1906. Solifluction, a component of sub-
    aerial denudation. *Journal of Geology,* 14, 91-112

Anderson, D.M., Reynolds, R.C. and Brown, J., 1969.
    Bentonite debris flows in northern Alaska. *Science,*
    164, 173-174

Anderson, D.M. and Morgenstern, N.R., 1973. Physics,
    chemistry and mechanics of frozen ground. *North
    American Contribution, 2nd International Permafrost
    Conference, Yukutsk,* (National Academy of Science,
    Washington) 256-288

Anderson D.M., Tice, A.R., and McKim, H.L., 1973. The
    unfrozen water and the apparent specific heat
    capacity of frozen soils. *North American
    Contribution, 2nd International Permafrost
    Conference, Yakutsk,* (National Academy of Science,
    Washington), 289-295

Anderson, D.M. and Tice, A.R., 1973. The unfrozen
    interfacial phase in frozen soil water systems.
    Ecological Studies, 4, 107-124

Anderson, H.W., 1947. Soil freezing and thawing as
    related to some vegetation, climatic, and soil
    variables. *Journal of Forestry,* 45, 94-101

Anderson, E.W. and Finlayson, B., 1975. Instruments for
    measuring soil creep. *British Geomorphological
    Research Group, Tech. Bull.* 16

Andrews, J.T., 1963. The analysis of frost heave data collected by B.H.J. Haywood from Shefferville, Labradore-Ungava. *Canadian Geographer,* 7, 163-173

Arakawa, K., 1966. Theoretical studies of ice segregation in soil. *Journal of Glaciology,* 6, 255-260

Arkell, W.J., 1943. The Pleistocene rocks at Trebetheric Point, north Cornwall: their interpretation and correlation. *Proceedings of the Geologists' Association,* 54, 141-170

Atkinson, H.B. and Bay, C.E., 1940. Some factors affecting frost penetration. *Transactions of the American Geophysical Union,* 3, 935-947

Atterberg, A., 1911. Uber die physikalish Bodenuntersuchung und uber die plastizitat der tone, *Internationale Mitteilungen für Bodenkunde,* 1, 10-43

Ball, D.F. and Goodier, R., 1968. Large sorted stone stripes in the Rhinog Mountains, North Wales. *Geografiska Annaler,* 50, 54-59

Ball, D.F. and Goodier, R., 1970. Morphology and distribution of features resulting from frost action in Snowdonia. *Field Studies,* 3, 193-218

Ball, D.F., Mew, G. and Macphee, W.S.G., 1969. Soils of Snowdon. *Field Studies,* 3, 69-107

Baranov, I.Y., 1959. Geographical distribution of seasonally frozen ground and permafrost. *General Geocryology, V.A. Obrucher Institute of Permafrost Studies, Academy of Science,* Pt. 1. Chapter 7, 193-219, N.R.C. Canada Technical Translation 1121

Barnett, D.M., 1966. Preliminary field investigations of movement on certain Arctic slope forms. *Geographical Bulletin,* 8, 377-382

Baulig, H., 1956. Peneplaines et pediplaines. *Société belge études geographie,* 25, 25-58

Baulig, H., 1957. Peneplains and pediplains. *Bulletin of the Geological Society of America,* 68, 913-930

Baver, L.D., 1956. *Soil Physics,* (John Wiley, New York)

Benedict, J.B., 1966. Radiocarbon dates from a stone-banked terrace in the Colorado Rocky Mountains, U.S.A. *Geografiska Annaler,* 48A, 24-31

Benedict, J.B., 1969. Microfabric of patterned ground. *Arctic and Alpine Research,* 1, 45-48

Benedict, J.B., 1970. Downslope soil movement in a Colorado alpine region: rates, processes and climatic significance. *Arctic and Alpine Research,* 2, 165-226

Benedict, J.B., 1976. Frost Creep and gelifluction features: A Review. *Quaternary Research,* 6, 55-76

Benninghoff, W.S., 1952. Interaction of vegetation and soil frost phenomena. *Arctic,* 5, 34-44

Berggren W.P., 1943. Prediction of temperature-distribution in frozen soils. *Transactions of the American Geophysical Union,* 3, 71-77

Beskow, G., 1935. Tjälbildningen och tjällyftningen med särskild hänsyn till vägar och jarnägar. *Sveriges Geologiska Undersökning,* Arsbok 26, Ser. C, No. 375

Beskow, G., 1947. Scandinavian soil frost research of the past decade. *Highway Research Board Proceedings 27th Annual Meeting,* 372-382

Bjerrum, L., and Simons, N.E., 1960. Comparison of shear strength characteristics of normally consolidated clays. *Proceedings American Society of Civil Engineers Research Conference on Shear Strength of Cohesive Soils,* 711-726

Blüthgen, J., 1970. Problems of definition and geographical differentiation of the Subarctic with special regard to northern Europe. *Ecology of the Subarctic Regions, Proceedings of Helsinki Symposium,* UNESCO, 11-33

Bouyoucos, G.J., 1913. An investigation of soil temperature and some factors influencing it. *Michigan Agricultural Experimental Station Technical Bulletin,* 17, 196

Bouyoncos, G.J., 1915. Effects of temperature on movement of water vapour and capillary moisture in soil. *Journal of Agricultural Research,* 5, 141-172

Bouyoucos, G.J., 1916. The freezing point method as a new means of measuring the concentration of soil solution directly in the soil. *Michigan Agricultural Research Station Technical Bulletin,* 24, 1-44

Bowen, D.Q., 1970. South-East and central South Wales. *The Glaciations of Wales,* ed. Lewis, C.A. (Longman, London), 197-228

Bowen, D.Q., 1974. The Quaternary of Wales, in *The Upper Palaeozoic and Post Palaeozoic Rocks of Wales,* ed Owen, T.R. (University of Wales, Cardiff).

British Standards Institution, 1975. *Methods of test for soils for civil engineering purposes,* B.S. 1377.

Brown, R.J.E., 1967. Comparison of permafrost conditions in Canada and the U.S.S.R. *Polar Record,* 13, 741-751

Brown, R.J.E., 1971. Characteristics of the active layer in the permafrost region of Canada. *Proceedings of seminar on the Permafrost Active Layer, National Research Council of Canada, Associate Committee on Geotechnical Research,* Technical Memo, 103, 1-7

Brown, R.J.E. and Péwé, T.L., 1973. Distribution of permafrost in North America and its relation to the environment: A Review. *North American Contribution, 2nd International Permafrost Conference, Yakutsk,* (National Academy of Science, Washington), 71-100

Brown, R.J.E. and Kupsch, W.O., 1974. *Permafrost Terminology,* National Research Council, Canada, 14274, 62

Burt, T.P. and Williams, P.J., 1976. Hydraulic conductivity in frozen soils. *Earth Surface Processes,* 1, 349-360

Cailleux, A. and Taylor, G., 1954. Cryopédologie, étude des sols gelés. *Expéditions Polaires Francaises, Missions Paul-Emile Victor IV Paris. Hermann et Gie, Actualitiés Scientifiques et Industrielles,* 1203, 218

Cailleux, A. and Tricart, J., 1950. Une type de solifluction les coultées bouluses. *Revue Géomorphologie dynamique,* 1, 4-46

Caine, T.N., 1963. Movement of low angle scree slopes in the Lake District, northern England. *Revue Géomorphologie dynamique,* 14, 171-177

Caine, T.N., 1968. The log-normal distribution and rates of soil movement: an example. *Revue Géomorphologie dynamique,* 18, 1-7

Carlson, H., 1952. Calculation of depth of thaw in frozen ground. *Highway Research Board, Special Report,* 2, 192-223

Carlson, H. and Kersten, M.S., 1953. Calculation of depth of freezing and thawing under pavements. *Highway Research Board, Bulletin,* 71, 81-98

Casegrande, A., 1931. Discussion of "A new theory of frost heaving" by A.C. Benkelman and F.R. Olmstead, *Highway Research Board Proceedings,* 11, 168-172

Casegrande, A., 1932. Research on the Atterberg Limits of soils. *Public Roads, Washington,* 13, 121-136

Chambers, M.J.G., 1966. Investigations of patterned ground at Signy Island, South Orkney Islands: I, Interpretation of mechanical analyses. *British Antarctic Survey Bulletin,* 9, 21-40

Chambers, M.J.G., 1966. Investigations of patterned ground at Signy Island, South Orkney Islands: II, thermal regimes in the active layer. *British Antarctic Survey Bulletin,* 10, 71-83

Chambers, M.J.G., 1967. Investigations of patterned ground at Signy Island, South Orkney Islands: III, miniature patterns, frost heaving and general conclusions. *British Antarctic Survey Bulletin,* 12, 1-22

Chandler, R.J., 1970a. The degredation of Lias Clay slopes in an area of the East Midlands. *Quarterly Journal of Engineering Geology,* 2, 161-181

Chandler, R.J., 1970b. A shallow slab slide in the Lias Clay near Uppingham, Rutland. *Geotechnique,* 20, 253-260

Chandler, R.J., 1972. Periglacial mudslides in Vestspitsbergen and their bearing on the origin of fossil 'solifluction' shears in low angled clay slopes. *Quarterly Journal of Engineering Geology,* 5, 223-241

Chandler, R.J, Kellaway, G.A., Skempton, A.W. and Wyatt, R.J., 1976. Valley slope sections in Jurassic strata near Bath, Somerset. *Philosophical Transactions of the Royal Society of London,* A. 283, 527-555

Cook, F.A., 1955. Near surface soil temperature measurements at Resolute Bay, North Western Territories. *Arctic,* 8, 237-249

Corte, A.E., 1962. Vertical migration of particles in front of a moving freezing plane. *Journal of Geophysical Research,* 67, 1085-1090

Costin, A.B., Thom, B.G., Wimbush, D.J., and Struiver, M., 1967. Nonsorted steps in the Mt. Kosciusko area, Australia. *Geological Society of America, Bull.,* 78, 979-999

Costin, A.B., 1972. Carbon-14 dates from the Snowy Mountains area, south-eastern Australia, and their interpretation. *Quarternary Research,* 2, 579-590

Costin, A.B., and Polach, H.A., 1972. Age and significance of slope deposits, Black Mountains, Canberra. *Australian Journal of Soil Research,* 11, 13-25

Crampton, C.B., 1965. An indurated horizon in soils of South Wales. *Journal of Soil Science,* 16, 230-241

Crampton, C.B., and Taylor, J.A., 1967. Solifluction terraces in South Wales. *Biuletyn Peryglacjalny,* 16, 15-36

Crawford, C.B., 1951. Soil Temperatures, a review of published records. *Highway Research Board Special Report,* 2, 17-40

Csathy, T.I. and Townsend, D.L., 1962. Pore size and field frost performance of soils. *Highway Research Board Bulletin,* 331, 67-80

Czeppe, Z., 1959. Uwagi oprocesie wymarzania glazow. *Czasopismo Geograficzne,* 30, 195-202

Czeppe, Z., 1960. Thermic differentiation of the active layer and its influence upon frost heave in periglacial regions (Spitsbergen). *Academie polonaise Sciences Bulletin Series Sci. Géographie,* 8, 149-152

Czudek, T. and Demek, J., 1973. The valley cryopediments in eastern Siberia. *Biuletyn Peryglacjalny,* 22, 117-130

Davison, C., 1889. On the creeping of the soil cap through the action of frost. *Geological Magazine,* 6, 255-261

De La Beche, H.T., 1839. Report on the geology of Cornwall, Devon and West Somerset. *Memoirs of the Geological Survey*

Dewey, H., Bromehead, C.E.N., Chatwin, C.P. and Dines, H.G., 1924. The geology of the country around Dartford. *Memoirs of the Geological Survey*

Dines, H.G., Hollingworth, S.E., Edwards, W., Buchan, S. and Welch, F.B., 1940. The mapping of head deposits. *Geological Magazine,* 77, 198-226

Dutkiewicz, L., 1967. The distribution of periglacial phenomena in N.W. Sorkapp, Spitsbergen. *Biuletyn Peryglacjalny,* 16, 37-84

Dylik, J., 1951.   Some periglacial structures in Pleistocene deposits of Middle Poland.   *Bulletin de la Societé des Sciences et des lettres de lodz,* 3, 1-6

Dylik, J., 1964.   Elements essentiels de la notion de periglaciaire.   *Biuletyn Peryglacjalny,* 14, 111-132

Dylik, J., 1967.   Solifluction, Congelefluxion and related slope processes.   *Geografiska Annaler,* 49, 167-177

Eden, R.A., Stevenson, I.P. and Edwards, W., 1957.   Geology of the country around Sheffield.   *Memoirs of the Geological Survey*

Edmonds, E.A., Wright, J.E., Beer K.E., Hawkes, J.R., Williams, M., Freshney, E.C. and Fenning, P.J., 1968.   Geology of the country around Okehampton.   *Memoirs of the Geological Survey*

Ellis, S., 1978a.   *An investigation of pedogenesis in an alpine-subalpine mountain environment, Okstindan, N. Norway,* unpublished Ph.D. thesis, University of Reading

Ellis, S., 1978b.   A study of the influence of ground movement on soil profile development in north-east Okstindan.   *Okstindan Research Project 1975 Preliminary Report,* Department of Geography, University of Reading, 38-62

Ellis, S., 1979.   Radiocarbon dating evidence for the initiation of solifluction ca. 5500 years B.P. at Okstindan, North Norway.   *Geografiska Annaler,* 61A, 29-33

Embleton, C. and King, C.A.M., 1975.   *Periglacial Geomorphology,* (Edward Arnold, London)

Evans, J.G., 1968.   Periglacial deposits in the Chalk of Wiltshire.   *Wiltshire Archaeological and Natural History Magazine,* 63, 12-27

Everett, D.H. and Hayes, J.M., 1965.   Capillary properties of some model pore systems with special reference to frost damage, *Revision Int. des Laboratories d'Essoir et de Reserches sur les Materiaux et Constructions Bulletin, New Series,* 27, 31-38

Everett, K.R., 1966.   Slope movement and related phenomena.   *Environment of the Cape Thompson Region, Alaska,* U.S. Atomic Energy Commission, Division of Technical Information, 175-220

Everett, K.R., 1967.   Mass-wasting in the Taseriaq area, West Greenland.   *Meddelelser om Grønland,* 165, 1-32

Fahey, B.D., 1973. An analysis of diurnal freeze-thaw cycles in the Indian Peaks region of the Colorado Front Range. *Arctic and Alpine Research,* 5, 269-281

Fahey, B.D., 1974. Seasonal frost heave and frost penetration measurements in the Indian Peaks Region of the Colorado Front Range. *Arctic and Alpine Research,* 6, 63-70

Fahey, B.D., 1979. Frost heaving of soils at two locations in southern Ontario, Canada. *Geoderma,* 22, 119-126

Flint, R.F. and Skinner, B.J., 1977. *Physical Geology,* (Wiley, New York), 594 p.

French, H.M., 1973. Cryopediments on the chalk of southern England. *Biuletyn Peryglacjalny,* 22, 149-156

French, H.M., 1974a. Mass-wasting at Sachs Harbour, Banks Island, N.W.T., Canada. *Arctic and Alpine Research,* 6, 71-78

French, H.M., 1974b. Active thermokarst processes eastern Banks Island, Western Canadian Arctic. *Canadian Journal of Earth Science,* 11, 785-794

French, H.M., 1976. *The Periglacial Environment,* (Longman, London)

French, H.M. and Egginton, P., 1973. Thermokarst development, Banks Island, Western Canadian Arctic. in *North American Contribution, 2nd International Permafrost Conference, Yakutsk,* (National Academy of Science, Washington), 2115, 203-212

Furrer, G., 1972. Bowegungsmedssungen auf solifluktionsdecken. *Zeitschrift für Geomorphologie,* Suppl, 13, 87-101

Furrer, G. and Bachman, F., 1972. Solifluctionsdecken im Schweizerischen Hochgebirge al Spiegel der postglazialen Landschaftsentwicklung. *Zeitschrift für Geomorphologie,* Suppl. 13, 163-172

Galloway, R.W., 1961a. Solifluction in Scotland. *Scottish Geographical Magazine,* 77, 75-87

Galloway, R.W., 1961b. Periglacial phenomena in Scotland. *Geografiska Annaler,* 43A, 348-353

Giles, R.S., 1973. Preliminary investigation of the periglacial environment in the Okstindan area. *Okstindan Research Project 1972 Preliminary Report,* Department of Geology, Univ. Reading, 35-42

Gold, L.W., 1957. Influence of snow cover on heat flow from the ground: some observations made in the Ottawa area. *Association of International Hydrological Science, Union Geodesique et Geophysique Internationale, Toronto,* 4, 13-21

Gold, L.W., Johnston, G.H., Slusarchuk, W.H. and Goodrich, L.E., 1972. Thermal effects in permafrost. *Proceedings of the Canadian Northern Pipeline Research Conference,* National Research Council Canada, Technical Memo, 104, 25-45

Goodrich, L.E., 1974. A one-dimensional numerical model for geothermal problems. *National Research Council of Canada, Division of Building Research,* Technical Paper 421

Gradwell, M.W., 1954. Soil frost studies at a high country station. *New Zealand Journal of Science and Technology* B, 36, 240-257

Graf, K., 1973. Vergleichende Betrachtungen zur solifluktion in Verschiedenen Breitenalgen. *Zeitschrift für Geomorphologie,* Suppl. 16, 104-154

Green, C.P. and Eden, M., 1973. Slope deposits on the weathered Dartmoor granite, England. *Zeitschrift für Geomorphologie,* Suppl. 18, 26-37

Harlan, R.L. and Nixon, J.F., 1978. Ground thermal regime, Chap. 3. *Geotechnical Engineering for Cold Regions,* ed Andersland, O.B. and Anderson, D.M., (McGraw-Hill, New York), 103-163

Harris, C., 1972a. Processes of soil movement in turf-banked solifluction lobes, Okstindan, Northern Norway. *Transactions of the Institute of British Geographers,* Special Publication 4, 155-174

Harris, C., 1972b. *Solifluction processes on a till slope in Okstindan, North Norway,* unpublished Ph.D. thesis, University of Reading

Harris, C., 1973a. Some factors affecting the rates and processes of periglacial mass movements. *Geografiska Annaler,* 55A, 24-28

Harris, C., 1973b. The Ice Age in Gower. *Gower,* 24, 74-79

Harris, C., 1974. Autumn, winter and spring soil temperatures in Okstindan, Norway. *Journal of Glaciology,* 13, 521-534

Harris, C., 1976. Zonation of periglacial features on the north-eastern flanks of Oksskolten. *Okstindan Research Project, 1974 Preliminary Report,* Department of Geography, University of Reading, 74-81

Harris, C., 1977. Engineering properties, groundwater conditions, and the nature of soil movement on a solifluction slope in North Norway. *Quarterly Journal of Engineering Geology*, 10, 27-43

Harris, C., 1980 (in press). Microstructures in solifluction sediments from South Wales and North Norway. *Biuletyn Peryglacjalny*, 28

Harris, C., (in preparation). Micromorphology of head deposits, South Wales and South-West England.

Harris, C. and Ellis, S., 1980. Micromorphology of soils in soliflucted materials, Okstindan, northern Norway. *Geoderma*, 23, 11-29

Harris, C. and Wright, M.D., 1980. Some last glaciation drift deposits near Pontypridd, South Wales. *Geological Journal*, 15, 7-20

Haywood, B.H.J., 1961. Studies of frost-heave cycles at Schefferville. *McGill Subarctic Research Papers*, 11, 6-10

Higashi, A., 1958. Experimental study of frost heaving. *U.S. Army Snow, Ice and Permafrost Research Establishment, Corps of Engineers*, Research Report 45

Higashi, A. and Corte, A.E., 1971. Solifluction: a model experiment. *Science*, 171, 480-482

Hill, J.B. and MacAlister, D.A., 1906. Geology of Falmouth and Truro. *Memoirs of the Geological Survey*

Ho, D.M., Harr, M.E. and Leonards, A., 1970. Transient temperature distribution in insulated pavements: predictions and observations. *Canadian Geotechnical Journal*, 7, 275-284

Holdgate, M.W., Allen, S.E. and Chambers, M.J.G., 1967. A preliminary investigation of the soils of Signy Island, South Orkney Islands. *British Antarctic Survey Bulletin*, 12, 53-71

Hollingworth, S.E., 1934. Some solifluction phenomena in the northern part of the Lake District. *Proceedings of the Geologist's Association*, 45, 167-188

Högbom, B., 1914. Uber die geologische Bedentung des Frostes. *Uppsala University, Geologisk Inst. Bulletin*, 12, 257-389

Horiguchi, H., 1979. Effect of the rate of heat removal on the rate of frost heaving. *Engineering Geology*, 13, 63-72

Hughes, O.L., 1972. Surficial geology and land classification, Mackenzie Valley Transportation Corridor. *Proceedings Canadian Northern Pipeline Research Conference,* National Research Council, Canada

Hustich, I., 1966. On the forest-tundra and the northern tree-lines. *Report of the Kevo Subarctic Research Station,* 3, 7-47

Hutchinson, J.N., 1974. Periglacial Solifluction: an approximate mechanism for clayey soils. *Géotechnique,* 24, 438-443

Hwang, C.T., Murray, D.W. and Brooker, E.W., 1972. A thermal analysis for structures in permafrost. *Canadian Geotechnical Journal,* 9, 33-46

Isaacs, R.M. and Code, J.A., 1972. Problems in engineering geology related to pipeline construction. *Proceedings Canadian Northern Pipeline Research Conference,* National Research Council, Canada, Technical Memo, 104, 147-178

Izotov, V.F., 1967. Pattern of soil freezing and thawing in waterlogged forests of the northern Taiga sub-zone. *Soviet Soil Science,* 6, 807-814

Jackson, K.A. and Chalmers, B., 1958. Freezing of liquids in porous media with special reference to frost heave in soils. *Journal of applied Physics,* 29, 1178-1181

Jahn, A., 1961. Quantitative analysis of some periglacial processes in Spitzbergen. *Universitet Wroclawski in Boleslawa Bieruta, zesvytynankowe, nanki przyrodnicze SerB.,* 5, 1-34

Jahn, A., 1976. Contemporaneous geomorphological processes in Longyeardalen, Vestspitsbergen (Svalbard). *Biuletyn Peryglacjalny,* 26, 253-268

Jahns, H.O., Miller, T.W., Power, L.D., Rickey, W.P., Taylor, T.P. and Wheeler, J.A., 1973. Permafrost protection for pipelines. *North American Contribution, 2nd International Permafrost Conference, Yakutsk,* (National Academy of Science, Washington), 673-684

James, P.A., 1971. The measurement of frost heave in the field. *British Geomorphological Research Group, Technical Bulletin,* 8

Jania, J., 1977. Debris forms on the skŏddefjellet slope. *Results of investigations of the Polish Scientific Spitzbergen Expeditions, 1970-1974 Vol. II, Acta. Universitats Wratislaviensis,* 387, 91-117

John, B.S., 1970. Pembrokeshire. in *The Glaciations of Wales*, ed Lewis, C.A., (Longman, London), 229-266

Jumikis, A.R., 1956. The soil freezing experiment. *Highway Research Board Bulletin*, 135, 150-165

Jumikis, A.R., 1973. Effect of porosity on amount of soil water transferred in a freezing silt. *North American Contribution, 2nd International Permafrost Conference, Yakutsk*, National Academy of Science, Washington), 305-310

Kallander, H., 1967. Patterned ground and solifluction at North Cape, Magerøy. *Lund Studies in Geography*, 40A, 24-40

Kaplar, C.W., 1968. New experiments to simplify frost susceptibility testing in soil. *Highway Research Record*, 215, 48-59

Kaplar, C.W., 1970. Phenomenon and mechanism of frost heaving. *Highway Research Record*, 304, 1-13

Kerfoot, D.E., 1969. *The geomorphology and permafrost conditions of Garry Island, N.W.T.*, unpublished Ph.D. thesis, University of British Columbia

Kerney, M.P., Brown, E.H. and Chandler, T.J., 1964. The late-glacial and post-glacial history of the Chalk escarpment near Brook, Kent. *Philosophical Transactions of the Royal Society of London*, B, 248, 135-204

Kersten, M.S., 1948. The thermal conductivity of soils. *Highway Research Board Proceedings*, 28

Kersten, M.S., 1949. Laboratory research for the determination of the thermal properties of soils. *University of Minnesota Engineering Experimental Station Bulletin*, 28, 161-166

Kersten, M.S., 1952. Thermal properties of soils. *Highway Research Board* special publ. 2, 161-166

Kersten, M.S., 1959. Frost penetration: Relationship to air temperatures and other factors. *Highway Research Board Bulletin*, 225, 45-80

King, C.A.M., 1976. *Northern England*, (Methuen, London)

King, R.B., 1972. Lobes in the Cairngorm Mountains, Scotland. *Biuletyn Peryglacjalny*, 21, 153-167

Kidson, C., 1971. The Quaternary history of the coasts of South West England, with special reference to the Bristol Channel coast. *Exeter Essays in Geography*, ed Gregory, K.J. and Ravenhill, W.L.D., Exeter, 1-22

Kirby, R.P., 1967. The fabric of head deposits in South Devon. *Proceedings of the Ussher Society,* 288-290

Kirkby, M.J., 1967. Measurement and theory of soil creep. *Journal of Geology,* 75, 359-378

Lachenbruch, A.H., 1970. Some estimates of the thermal effects of a heated pipeline in permafrost. *U.S. Geological Survey, Circular,* 632

Lamothe, L. and St. Onge, D., 1961. A note on periglacial erosion processes in the Isachsen Area, N.W.T. *Geographical Bulletin,* 16, 104-113

Lewis, C.A., 1970. The Upper Wye and Usk Regions. in *The Glaciations of Wales,* ed. Lewis, C.A., (Longman, London) 147-174

Lewis, C.A. and Lass, G.M., 1965. The drift terraces of Slaettaratindur, the Faeroes. *Geographical Journal,* 131, 247-253

Lock, G.S.H., Gunderson, J.R., Quon, D. and Donnelly, J.K., 1969. A study of one-dimensional ice formation with particular reference to periodic growth and decay. *International Journal of Heat Mass Transfer,* 12, 1343-1352

Lozinski, W., 1912. Die periglazial fazies der mechanischen verwitterung. *11th International Geological Congress,* Stockholm 1910, 2, 1039-1053

Lundqvist, G., 1949. The orientation of the block material in certain species of flow earth. *Geografiska Annaler,* 31, 335-347

Lundqvist, J., 1962. Patterned ground and related phenomena in Sweden. *Svergies Geologiska Undersökning,* Arsbok 55(7), 4-101

Mackay, J.R., 1966. Segregated epigenetic ice and slumps in permafrost, Mackenzie Delta, N.W.T. *Geographical Bulletin,* 8, 59-80

Mackay, J.R., 1972. The world of underground ice. *Annals of the Association of American Geographers,* 62, 1-22

Mackay, J.R. and Matthews, W.H., 1972. Geomorphology and Quarternary history of the Mackenzie River Valley near Fort Good Hope, N.W.T., Canada. *Canadian Journal of Earth Science,* 10, 26-41

Matthews, B., 1967. Automatic measurement of frost heave, results from Malham and Rodley (Yorkshire). *Geoderma,* 1, 107-115

Matthews, J.A., 1980. Some problems and implications of ^{14}C dates from a podzol buried beneath an end moraine at Haugabreen, southern Norway. *Geografiska Annaler,* 62A, 185-208.

McGaw, R., 1972. Frost heaving versus depth to water table. *Highway Research Record,* 395, 45-55

McKeown, M.C., Edmonds, E.A., Williams, M., Freshney, E.C. and Masson-Smith, D.J., 1973. Geology of the country around Boscastle and Holsworthy. *Memoirs of the Geological Survey*

McRoberts, E.C., 1973. *The stability of slopes in permafrost.* unpublished Ph.D thesis, University of Alberta, Edmonton

McRoberts, E.C., 1978. Slope stability in cold regions. in *Geotechnical Engineering for Cold Regions,* ed. Andersland, O.B. and Anderson, D.M., (McGraw-Hill, New York), 363-404

McRoberts, E.C. and Morgenstern, N.R., 1974a. The stability of thawing slopes. *Canadian Geotechnical Journal,* 11, 447-469

McRoberts, E.C. and Morgenstern, N.R., 1974b. The stability of slopes in frozen soil, Mackenzie Valley, N.W.T. *Canadian Geotechnical Journal,* 11, 554-573

Miller, R.D., 1972. Freezing and heaving of saturated and unsaturated soils. *Highway Research Board,* 393, 1-11

Morgenstern, N.R. and Nixon, J.F., 1971. One dimensional consolidation of thawing soils. *Canadian Geotechnical Journal,* 8, 558-565

Mottershead, D.N., 1971. Coastal head deposits between Start Point and Hope Cove, Devon. *Field Studies,* 5, 433-453

Mottershead, D.N., 1977. *South-West England,* Guidebook for Excursions A6 and C6, INQUA, 10th Congress, 1977

Mottershead, D.N., 1978. High altitude solifluction and post-glacial vegetation, Arkle, Sutherland. *Transactions of the Botanical Society of Edinburgh,* 43, 17-24

Mottershead, D.N. and White, I.D., 1969. Some solifluction terraces in Sutherland. *Transactions of the Botanical Society of Edinburgh,* 40, 604-620

Nicholson, F.H. and Granberg, H.B., 1973. Permafrost and snow cover relationships near Schefferville. *North American Contribution, 2nd International Permafrost Conference, Yakutsk,* (National Academy of Science, Washington), 151-158

Nixon, J.F. and McRoberts, E.C., 1973. A study of some factors affecting the thawing of frozen soils. *Canadian Geotechnical Journal,* 10, 439-452

Nixon, J.F. and Morgenstern, N.R., 1973. The residual stress in thawing soils. *Canadian Geotechnical Journal,* 10, 617-631

Nixon, J.F. and Ladanyi, B., 1978. Thaw consolidation. in *Geotechnical Engineering for Cold Regions,* ed. Andersland, D.B. and Anderson, D.M., (McGraw-Hill, New York), 164-215

Østrem, G., 1965. Problems of dating ice-cored moraines. *Geografiska Annaler,* 47A, 1-38

Outcalt, S.I., 1969. Weather and diurnal frozen soil structure at Charlottesville, Virginia. *Water Resources Research,* 5, 1377-1381

Outcalt, S.I., 1970. Study of time dependence during serial needle ice events. *Archiv für Meteorologie, Geophysik und Bioklimatologie,* A, 19, 329-337

Outcalt, S.I., 1972. The development and application of a simple digital surface climatic simulator. *Journal of Applied Meteorology,* 11, 629-656

Outcalt, S.I., 1973. A simulation sensitivity analysis of the needle ice growth environment. *North American Contribution, 2nd International Permafrost Conference, Yakutsk,* (National Academy of Science, Washington), 228-234

Outcalt, S.I., Goodwin, C., Weller, G. and Brown, J., 1975. Computer simulation of the snowmelt and soil thermal regime at Barrow, Alaska. *Water Resources Research,* 11, 709-715

Palzell, D. and Durrance, E.M., 1980. The evolution of the Valley of Rocks, North Devon. *Transactions of the Institute of Geographers,* New Series, 5, 66-79

Penner, E., 1959. The mechanics of frost heaving in soils. *Highway Research Board Bulletin,* 225, 1-22

Penner, E., 1968. Particle size as a basis for predicting frost action in soils. *Soils and Foundations,* 8, 21-29

Penner, E., 1970a. Frost heaving forces in Leda clay. *Canadian Geotechnical Journal,* 7, 8-16

Penner, E., 1970b. Thermal conductivity of frozen soils. *Canadian Journal of Earth Science,* 7, 982-987

Penner, E. 1972. Influence of freezing rate on frost heaving. *Highway Research Record,* 393, 56-64

Péwé, T.L., 1969. The periglacial environment. in *The Periglacial Environment,* ed. Péwé, T.L., (McGill-Queens Univ. Press, Montreal), 1-11

Péwé, T.L., 1975. Quaternary Geology of Alaska. *U.S. Geological Survey Professional Paper,* 835, 145

Pihlainen, J.A., 1962. Inuvik, North West Territories - Engineering site information. *Division of Building Research, National Research Council of Canada, Technical Paper,* 135, 18

Pissart, A., 1977. Apparition et évolution des sols structuraux périglacaires de haute montagne. Experiences de terrain du Chambeyron (Alps, France). in *Formain Formengesellschaften und Untergrenzen in des heutigen periglazialen Höhenstufen des Hochgebirge Europas und Afrikas zwischen Arktis und Aquator,* ed. Poser, H., Abhandlungen der Akademie der wissenschagten in Gottingen Nr 31, 142-156

Potts, A.S., 1971. Fossil cryonival features in Central Wales. *Geografiska Annaler,* 53A, 39-51

Price, L.W., 1969. The collapse of solifluction lobes as a factor in vegetating blockfields. *Arctic,* 22, 395-402

Price, L.W., 1970. *Morphology and ecology of solifluction lobe development - Ruby Range, Yukon Territory,* Unpublished Ph.D. thesis, University of Illinois, Urbana

Price, L.W., 1971. Vegetation, microtopography and depth of active layer on different exposures in subarctic alpine tundra. *Ecology,* 52(4), 638-647

Price, L.W., 1972. The periglacial environment, permafrost and man. *Resource Paper 14, Association of American Geographers Committee on College Geography,* 88

Price, L.W., 1973. Rates of mass-wasting in the Ruby Range, Yukon Territory. *North American Contribution, 2nd International Permafrost Conference, Yakutsk,* (National Academy of Science, Washington), 235-245

Ragg, J.M. and Bibby, J.S., 1966. Frost weathering and solifluction products in southern Scotland. *Geografiska Annaler,* 48A, 12-23

Rapp, A., 1960. Recent development of mountain slopes in Karkevagge and surroundings, northern Sweden. *Geografiska Annaler,* 42A, 71-200

Rapp, A., 1962. Karkevagge, some recordings of mass movements in the Northern Scandinavian Mountains. *Biuletyn Peryglacjalny,* 11, 287-309

Raup, H.M., 1951. Vegetation and cryoplanation. *Ohio Journal of Science,* 51(3), 105-116

Rein, R.G. and Burrous, C.M., 1980. Laboratory measurements of subsurface displacements during thaw of low-angle slopes of a frost susceptible soil. *Arctic and Alpine Research,* 12, 349-358

Rose, J., 1980. The western side of Loch Lomond. in *Quaternary Research Association Field Guide, Glasgow Region,* 37-39

Rudberg, S., 1958. Some observations concerning mass movements on slopes in Sweden. *Geologiska föreningens i Stockholm forhandlingar,* 80, 114-125

Rudberg, S., 1962. A report on some field observations concerning periglacial geomorphology and mass movements on slopes in Sweden. *Biuletyn Peryglacjalny,* 11, 311-323

Rudberg, S., 1964. Slow mass movement processes and slope development in the Norra Storfjall area, southern Swedish Lappland. *Zeitschrift für Geomorphologie,* Suppl. 5, 192-203

Rune, O., 1965. The mountain regions of Lappland. *Acta Phytogeographica Suecica,* 50, 64-77

Shakesby, R.A., 1975. An investigation into the origin of the deposits in a chalk dry valley of the South Downs, Southern England. *Univ. Edinburgh Dept. Geography Research Discussion Paper,* 5, 25

Sharp, R.P., 1942. Soil structures in the St. Elias Range, Yukon Territory. *Journal of Geomorphology,* 5, 274-301

Sharpe, C.F.S., 1938. *Landslides and Related Phenomena,* (Columbia Univ. Press, New York)

Sharpenseel, H.W. and Schiffmann, H., 1977. Radiocarbon dating of soils, a review. *Zeitschrift für Pflanzenernährung Düngung und Bodenkunde,* 140, 159-174

Shilts, W.N., 1974. Physical and chemical properties of unconsolidated sediments in permanently frozen terrain, District of Keewatin. *Geological Survey of Canada,* Paper 74, 229-235

Sigafoos, R.S. and Hopkins, D.M., 1952. Soil stability on slopes in regions of perennially frozen ground. *Highway Research Board Special Report* 2, 176-192

Siple, P.A., 1952. Ice blocked drainage as a principle factor in frost heave, slump and solifluction. *Highway Research Board Special Report* 2, 172-175

Sissons, J.B., 1977. The loch Lomond readvance in the northern mainland of Scotland. *Studies in the Scottish Lateglacial Environment,* ed. Gray, J.M. and Lowe, J.J., (Pergamon Press), 45-60

Skempton, A.W., 1957. Discussion: the planning and design of the new Hong Kong airport. *Proceedings of the Institute of Civil Engineers,* 7, 305-307

Skempton, A.W., 1964. Long term stability of clay slopes. *Géotechnique,* 14, 77-101

Skempton, A.W. and De Lory, F.A., 1957. Stability of natural slopes in London Clay. *Proceedings of the 4th International Conference on Soil Mechanics,* 2, 378-381

Skempton, A.W. and Weeks, A.G., 1976. The Quaternary history of the Lower Greensand escarpment and Weald clay vale near Sevenoaks, Kent. *Philosophical Transactions of the Royal Society of London,* A. 283, 493-526

Skempton, A.W. and Petley, D.J., 1967. The strength along structural discontinuities in stiff clays. *Proceedings Geotechnical Conference, Oslo,* 2, 29-46

Small, R.J., Clark, M.J. and Lewin, J., 1970. The periglacial rock-stream at Clatford Bottom, Marlborough Downs Wiltshire. *Proceedings of the Geologists' Association,* 81, 87-98

Smith, J. 1956. Some moving soils in Spitsbergen. *Journal of Soil Science,* 7, 10-21

Smith, M.V. and Tvede, A., 1977. The computer simulation of frost penetration beneath highways. *Canadian Geotechnical Journal,* 14, 167-179

Smith, W.O., 1939. Thermal conductivity of moist soils. *Proceedings of the Soil Science Society of America,* 4

Stephan, J., 1890. Uber die Theorie der Eisbildung, isbesonders Uber die Eisbildung im Polarmere. *Sitzungsberichte der Mathematisch - Naturwissenschaftliche Classe der Kaiserlichen Alcademie der Wissenschaften,* Wein XCVIII, IIa, 965-983

Stephens, N., 1961. Pleistocene events in North Devon. *Proceedings of the Geologists' Association,* 72, 469-472

Stephens, N., 1970. The west country and southern Ireland. in *The Glaciations of Wales,* ed. Lewis, C.A., (Longman, London), 267-314

Suess, H.E., 1970. Bristlecone-pine calibration of the radiocarbon time-scale 5200 BC to the present. in *Radiocarbon Variations and Absolute Chronology,* ed. Olsson, I.U., (Almqvist and Wiksell, Stockholm), 303-311

Sutherland, H.B. and Gaskin, P.N., 1973. Pore water and heaving pressure developed in partially frozen soils. *North American Contribution 2nd International Permafrost Conference, Yakutsk,* (National Academy of Science, Washington), 409-418

Svenovus, F.V., 1904. Yttrande med anledning av R. Sernanders föredrag: Några bidrag till de centralskandinaviska fjalltrakternas postglaciala geologi. *Geologiska föreningens Stockholm förhandlingar,* 26

Taber, S., 1929. Frost heaving. *Journal of Geology,* 37, 428-461

Taber, S., 1930. The mechanics of frost heaving. *Journal of Geology,* 38, 303-317

Taber, S., 1943. Perennially frozen ground in Alaska: its origin and history. *Bulletin of the Geological Society of America,* 54, 1433-1548

Takagi, S., 1978. Segregation freezing as the cause of suction force for ice lens formation. *Cold Regions and Engineering Laboratory* Report 78, Honover, New Hampshire

Takagi, S., 1979. Segregation freezing as the cause of suction force for ice lens formation. *Engineering Geology,* 13, 93-100

Tivy, J., 1962. An investigation of certain slope deposits in the Lowther Hills, Southern Uplands of Scotland. *Transactions of the Institute of British Geographers,* 30, 59-73

Tricart, J., 1970. *Geomorphology of Cold Environments,* (Macmillan, London)

Troll, C., 1944. Strukturboden, Solifluktion und frostklimate de erde. *Geologische Rundschan,* 34, 545-694

Troll, C., 1958. Structure soils solifluction and frost climates of the world. *U.S. Army Snow, Ice and Permafrost Research Establishment, Corps of Engineers, Wilmette* III, Translation 43, 121

Tufnel, L., 1969. The range of periglacial phenomena in northern England. *Biuletyn Peryglacjalny,* 19, 291-323

Van Rooyen, M. and Winterkorn, H.F., 1957. Theoretical and practical aspects of the thermal conductivity of soils and similar granular systems. *Highway Research Board Bulletin,* 168, 143-205

Varnes, A.J., 1958. Landslide types and processes. in *Landslides and Engineering Practice,* ed. Eckert, E.B., Highway Research Board Special Report, 29, 20-45

Walter, H. and Leith, H., 1967. *Klimadiagramm-Weltatlas,* (VEB Gustav Fischer Verlag, Jena)

Washburn, A.L., 1967. Instrumental observations of mass-wasting in the Mesters Vig district, N.E. Greenland. *Meddelelser om Grønland,* 166, 1-297

Washburn, A.L., 1969. Weathering, frost action and patterned ground in the Mesters Vig District, Northeast Greenland. *Meddelelser om Grønland,* 176, 303

Washburn, A.L., 1973. *Periglacial Processes and Environments,* (Edward Arnold, London)

Washburn, A.L., 1979. *Geocryology,* (Edward Arnold, London)

Waters, R.S., 1964. The Pleistocene legacy to the geomorphology of Dartmoor. *Dartmoor Essays,* ed. Simmons, I.G., (University of Exeter, Exeter) 73-96

Watson, E., 1966. Two nivation cirques near Aberystwyth, Wales. *Biuletyn Peryglacjalny,* 15, 79-101

Watson, E. 1970. The Cardigan Bay area. in *The Glaciations of Wales,* ed. Lewis, C.A., (Longman, London), 125-145

Watson, E. and Watson, S., 1967. The periglacial origin of the drifts at Morfa-bychan, near Aberystwyth. *Geological Journal*, 5, 419-440

Watson, E. and Watson S., 1970. The coastal periglacial slope deposits of the Cotentin Peninsula. *Transactions of the Institute of British Geographers*, 49, 125-144

Watson, E., 1976. Field excursions in the Aberystwyth region 1-10 July, 1975. *Biuletyn Peryglacjalny*, 26, 79-112

Weeks, A.G., 1969. The stability of slopes in south-east England as affected by periglacial activity. *Quarterly Journal of Engineering Geology*, 5, 223-241

Whittow, J.B. and Ball, D.F., 1970. North-west Wales. in *The Glaciations of Wales*, ed. Lewis, C.A., (Longman, London)

Williams, P.J., 1957. Some investigations into solifluction features in Norway. *Geographical Journal*, 123, 42-58

Williams, P.J., 1961. Climatic factors controlling the distribution of certain frozen ground phenomena. *Geografiska Annaler*, 43A, 339-347

Williams, P.J., 1964. Experimental determination of apparent specific heats of frozen soils. *Géotechnique*, 14, 133-142

Williams, P.J., 1966. Downslope soil movement at a Sub-Arctic location with regard to variations in depth. *Canadian Geotechnical Journal*, 3, 191-203

Williams, P.J., 1968a. Properties and behaviour of freezing soils. *Division of Building Research, National Research Council, Canada*, Research Paper 359

Williams, P.J., 1968b. Ice distribution in permafrost profiles. *Canadian Journal of Earth Science*, 5, 1381-1386

Williams, P.J., 1972. Use of the ice-water surface tension concept in engineering practice. *Highway Research Record*, 93, 19-29

Williams, P.J., 1977. Thermodynamic conditions for ice accumulation in freezing soils. *Proceedings International Symposium, Frost Action in Soils, Luleå, Sweden*, 1, 42-53

Williams, P.J. and Nickling, W.G., 1971. Ground thermal regime in cold regions. *2nd Geulph Symposium on Geomorphology,* 27-43

Winterkorn, H. and Baver, L.D., 1934. Sorption of liquids by soil colloids: 1, liquid intake and swelling by soil colloid material. *Soil Science,* 38, 291-298

Wood, A., 1959. The erosional history of the cliffs around Aberystwyth. *Liverpool and Manchester geological Journal,* 2, 271

Wright, M.D. and Harris, C., 1980. Superficial deposits in the South Wales Coalfield. *Cliff and Slope Stability in South Wales,* ed. Perkins, J., Department of Extra-Mural Studies, University College, Cardiff, 193-205

Wright, M.D., (in preparation). A study of Pleistocene drift deposits in the South Wales Coalfields, Ph.D., Open University

Young, R.N. and Osler, J.C., 1971. Heave and heaving pressures in frozen soils. *Canadian Geotechnical Journal,* 8, 272-282

Mackenzie Delta, N.W.T., Canada, 68, 145
Mackenzie River Valley, N.W.T., Canada, 108, 142, 144, 145
Marlborough Downs, England, 156
Mass-wasting, discussion and definitions, 1, 5
Mesters Vig, Greenland, 76-79, 104, 108, 116, 124
Micaceous sediments and solifluction processes, 107
Moisture migration in freezing soils, 19, 27
Morfa-bychan, Dyfed, Wales, 164, 170-171
Mudflows, 3, 14, 143-144
Mt. Chavagl, Swiss Alps, 114
Mynydd Bach, Wales, 164
Mynydd Eppynt, Wales, 164

Needle ice, 74, 75, 82, 83-85
Neuman equation, 40, 43, 45, fig.22
Nivation hollows, 165
Niwot Ridge, Colorado Rockies, U.S.A., 7, 70-72, 79-80
Nonsorted sheet, 126
Norra Storfjäll, Sweden, 112, 114
North American periglacial climates, 9, fig.2
North Downs, England, 155
North York Moors, England, 153
Northern England, 171-172
Northfork, California, U.S.A., 30
Norway, 14, 104, 127, 128, 129, see also
    Okstindan
Nottingham, England, 75

Okehampton, Devon, 157
Okstindan, Norway, 14, 29, 36, 70, 80, 108, 112, 114, 119,
    125, 127, 128, 131, 132, 136, fig.7
Oxfordshire, England, 149
Oxwich Bay, Gower, Wales, 170

Pararctic zone, 8, fig.1
Paraboreal zone, 8, fig.1
Pennines, England, 172
Pembrokeshire, Wales, 163
Periglacial environments, 5-16
Periglacial mass wasting, see *Mass wasting*
Permafrost, 3, 4, 5, 6, 9, 15, see also *Ice-rich permafrost*
Permeability, see *Hydraulic conductivity*
Permian bedrock, 158
Petersfield, Hants, England, 157
Physiognomic forest line, 12
Plastic limit, see *Atterberg Limits*
Plasticity Index, 88-91
Plastic tubing for measuring soil movement, 109
Pleistocene periglacial slope deposits, Britain, 148-176
Porewater pressure during freeze, 53, fig.29
Porewater pressure during thaw, 96-103, 122, 153, 154, fig.52
Porewater suction, 61, fig.33
Port Eynon, Gower, Wales, 170
Potential frost creep, 76